The Athletic Brain

Many parts of the athlete's body are important for performance, such as strong muscles and bones; healthy lungs and heart; and several sensory systems, including the vision, touch, and joint position senses, and the vestibular system for balance. However, the critical element for athletic performance is "not what you have but how you use it." The organ that decides "how you use it" is the brain. This book explains how the brain allows the athlete to perform.

The book starts with brief review of brain anatomy, which is necessary to understand how the brain functions. The book then outlines many critical aspects of the athlete's brain, including learning motor skills; decision-making; hand preference; visual perception of speed, distance, and direction; balance; emotions and moods. Finally, the book discusses the adverse influence of brain injuries.

Kenneth M. Heilman, MD, is currently a Distinguished Professor Emeritus of Neurology at the University of Florida and a Staff Neurologist at the Veterans Affairs Hospital in Gainesville, Florida. In addition to being a clinical neurologist and an educator, he performs research on brain functions and diseases. He has written more than 600 journal articles and authored or edited 16 books. He has received many honors including being an Honorary Member of the American Neurological Association and a Fellow of the Academy of Neurology.

The Athletic Brain

Kenneth M. Heilman

NEW YORK AND LONDON

First published 2018
by Routledge
711 Third Avenue, New York, NY 10017

and by Routledge
2 Park Square, Milton Park, Abingdon, Oxon, OX14 4RN

Routledge is an imprint of the Taylor & Francis Group, an informa business

© 2018 Taylor & Francis

The right of Kenneth M. Heilman to be identified as author of this work has been asserted by him in accordance with sections 77 and 78 of the Copyright, Designs and Patents Act 1988.

All rights reserved. No part of this book may be reprinted or reproduced or utilised in any form or by any electronic, mechanical, or other means, now known or hereafter invented, including photocopying and recording, or in any information storage or retrieval system, without permission in writing from the publishers.

Trademark notice: Product or corporate names may be trademarks or registered trademarks, and are used only for identification and explanation without intent to infringe.

Library of Congress Cataloging-in-Publication Data
A catalog record for this title has been requested

ISBN: 978-1-138-54216-7 (hbk)
ISBN: 978-1-138-54217-4 (pbk)
ISBN: 978-0-429-42802-9 (ebk)

Typeset in Times New Roman
by codeMantra

This book is dedicated to my brother Fred, who taught me to love sports.

Contents

List of Figures   ix

1 Introduction   1
  Brain Anatomy 1
    Brief Overview 1
    Sensory Cortex and Sensory Association Areas 4
    Motor and Premotor Cortex 7
    Prefrontal Cortex 8
    Subcortical Areas 8
    Cerebellum 9
  Types of Sports 11

2 Movement Action Programming   13
  The Praxis "How" System 13
  Learning New Athletic Motor Skills 24
    Motor Memory 24
    Practice 30
    Specialization 30
    Precision 30
  Action Sequencing 33
  Action Recognition 34
  Strength 37

3 Action-Intention: "When" Programming   39
  Introduction 39
  Action Initiation 41
  Deficits in Planning and Action Initiation: Executive Dysfunction 42
    Akinesia and Abulia 42
    Planning 45
    Hypokinesia 45
  Reward 46
    Reward Network 46
    Reinforcement 47
  Response Inhibition 49

Motor (Action) Impersistence 50
Motor Perseveration 51

4 Handedness   53

5 Attention   63
Definition 63
Neuroanatomy of Attentional Networks 64
Vigilance 66

6 Visual Perception   70

7 Balance   74

8 Emotions and Mood   77
Anger 77
Depression 79

9 Traumatic Brain Injury   81
Memory 82
Executive Functions 83
Emotions and Mood 84
Chronic Traumatic Encephalopathy 85

10 Influence of Exercise on Age Related Cognitive Decline and Dementia   87

References   89
Index   95

Figures

1.1	Cartoon of a neuron making a synapse with another neuron. This figure illustrates the dendrites, axons, and cell body	2
1.2	Image of the left hemisphere as seen from the left side of the body. The brain has four major lobes: frontal, parietal, temporal, and occipital	3
1.3	Somatosensory cortex	4
1.4	Image of left hemisphere, illustrating the primary visual, somatosensory, and auditory areas	5
1.5	Dorsal "where" and ventral "what" visual networks	6
1.6	Motor and premotor cortices, including convexity and supplementary motor area	7
1.7	Corpus callosum, connecting the right and left hemispheres	9
1.8	Cartoon of the structures deep in the right and left hemispheres, including the basal ganglia (caudate, putamen, and globus pallidus) as well as the thalamus and the internal capsule	10
1.9	Cerebellum	10
2.1	Motor cortex, corticospinal tract in brain and spinal cord	14
2.2	Supramarginal and angular gyri of the inferior parietal lobe	19
2.3	Superior parietal lobe	21
2.4	Superior longitudinal fasciculus	22
2.5	Praxis network	23
2.6	Hippocampus	24
2.7	Right fusiform gyrus	35
3.1	Wernicke's area	41
3.2	Nucleus accumbens	47
3.3	Substantial nigra and ventral tegmental area	48
5.1	Prefrontal connectivity	64
5.2	Mesencephalic reticular formation	67
5.3	Yerkes-Dobson curve	68
5.4	Corbetta's two attentional pathways	69

x *Figures*

6.1	Visual pathways from the retina to primary visual cortex	71
6.2	Cortical areas MT and MST	73
7.1	Pedunculopontine nucleus	76
8.1	Amygdala	78
9.1	Orbitofrontal cortex	84

1 Introduction

Many organs of the athlete's body are important for their athletic performance, including healthy joints and strong bones. In addition, the successful athlete must use several sensory systems. In most sports, taste and smell are not very important, and while hearing is often needed as a means of communication, it is not critical. The most important senses are vision, touch, and joint position sense. The vestibular system, which is in our inner ears, is also important for sensing gravity and spatial movements. The successful athlete also needs to be able to move their body, and therefore muscles are also important. These muscles need much energy, and therefore, adequate oxygen and glucose must be available. However, the old saying "It is not what you have but how you use it" is probably more correct for athletes than for most other human endeavors, and the organ that decides "how to use it" is the human brain. This book is written about how the brain allows the athlete to perform.

The majority of people reading this book will probably not be neuroscientists, neurologists, or neurosurgeons. However, in order to understand how the brain mediates athletic abilities, we need to briefly describe the anatomy of the brain.

Brain Anatomy

Brief Overview

The brain contains millions of cells called neurons (Figure 1.1). These neurons usually have three major divisions: the cell body, the axon, and the dendrites. The cell body is the metabolic center of the neuron, supplying energy to the entire neuron. The dendrites are small branches of the neuron that emanate from the cell body, and the axon is like a long tail. The dendrites from one neuron meet with dendrites from other neurons that are close neighbors. In contrast, the long axons of neurons meet the dendrites or axons of other neurons that are distant. The meeting place between neurons is called a synapse (Figure 1.1).

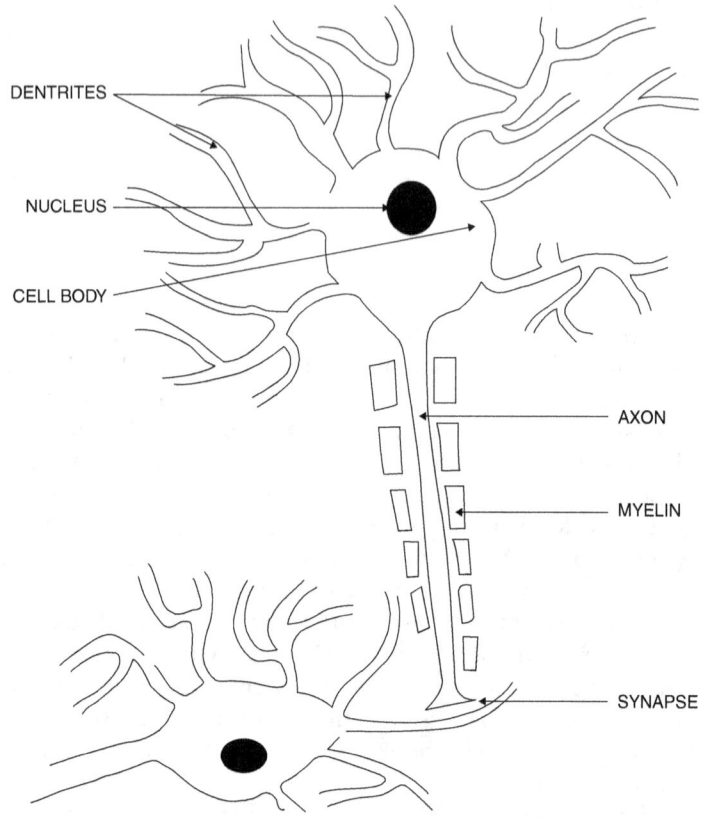

Figure 1.1 **Neuron and synapse**. This figure illustrates the dendrites, axons, cell body, and a synapse. The nerve cell or neuron has three major components: the cell body, the dendrites, and the axon. The cell body of the neuron contains the apparatus for producing energy from the oxygen and glucose it receives from the blood. This energy is important to keep an electrical charge across the cell membrane by moving electrolytes across the membrane. The nucleus contains the DNA that is important for the production of proteins. The dendrites are like branches of a tree that meet other neurons and communicate with these neurons. The axon is like a long branch of a tree that also communicates with other neurons. The place where a neuron meets another neuron or muscle is called a synapse. At the end of the neuron, there are chemicals that are given off when the neuron fires and can either excite or inhibit the neuron which the dendrite or axon meets.

Neurons build up electric charges, and when they are activated, they discharge and send these electrical charges down the dendrites and axons like an electrical cable. When the current reaches the end of these branches, where they meet the ends of other neurons (synapses), they

Introduction 3

give off chemicals called neurotransmitters. The chemical neurotransmitters can either help to activate/excite the neuron that they are meeting at the synapse or they can inhibit this neuron. When they excite a neuron, they can cause this neuron to discharge (fire) or make it more likely to fire. Alternatively, they can inhibit this neuron from firing.

There are many different estimates of the number of neurons in the brain, and these range from about 20 to 80 billion. It has been estimated that there are more than 100 trillion connections between these neurons. Not all neurons are connected to each other, but when neurons are connected, they form a network, and different networks in the human brain perform different functions. Some of these networks store information. Others analyze incoming sensory information as well as program movements. There are other neuronal networks that monitor our body, such as our oxygen and glucose levels, as well as other networks that help control organs of our body, such as the heart.

In humans, most of the neurons are on the surface of the brain, and these neurons help form the cerebral cortex. The brain's cerebral cortex is divided into two hemispheres: one hemisphere on the right and the other on the left. Each hemisphere is divided into four major sections or lobes, including the frontal, parietal, temporal, and occipital lobes (Figure 1.2).

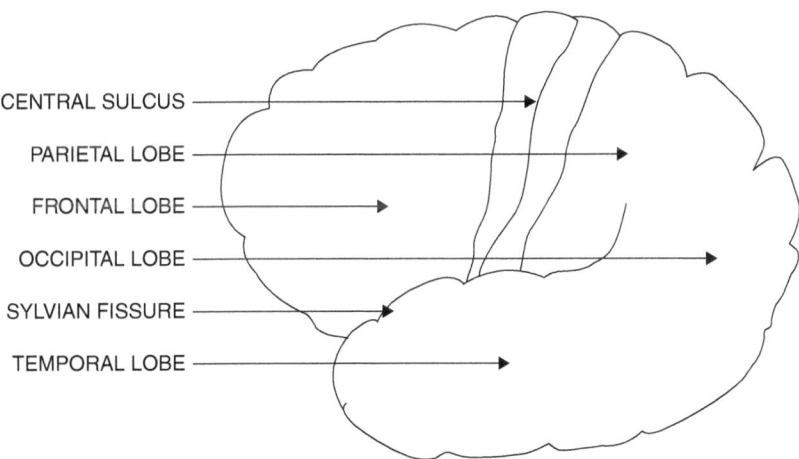

Figure 1.2 **Major lobes of the brain.** Image of the left hemisphere as seen from the left side. Each hemisphere contains four major lobes. In the front is the frontal lobe, and in the back is the occipital lobe. Between these lobes on the bottom is the temporal lobe, and above the back part of the temporal lobe is the parietal lobe. The frontal lobe is separated from the parietal lobe by the central sulcus, and the temporal lobe is separated from the parietal and frontal lobes by the Sylvian fissure.

4 Introduction

Sensory Cortex and Sensory Association Areas

The cerebral cortex receives sensory information in areas called primary sensory cortex. Visual stimuli come to the occipital cortex, at the back of the brain (Figure 1.2). Touch, temperature, pain, and joint position come to the somatosensory cortex, which is in the most anterior (front) portion of the parietal lobe (Figure 1.3). The primary auditory cortex is at the very top of the temporal lobe (Figure 1.4).

Each of these primary areas performs an analysis of the incoming information. For example, when sounds come into the ear, they are changed to electrical signals and sent to the primary auditory cortex. The primary auditory cortex determines these sounds' frequencies (pitch), amplitudes (loudness), and durations. Similarly, with vision, each half of the retina goes to the primary visual cortex in the occipital lobe that is on the same side, and since each half of the retina detects light coming in from the opposite side, the left primary visual in the occipital lobe area detects stimuli in the right side of space, and the right primary visual area detects stimuli in the left side of space. The primary visual cortex then determines the location, orientation, size (e.g., length), color, and movement

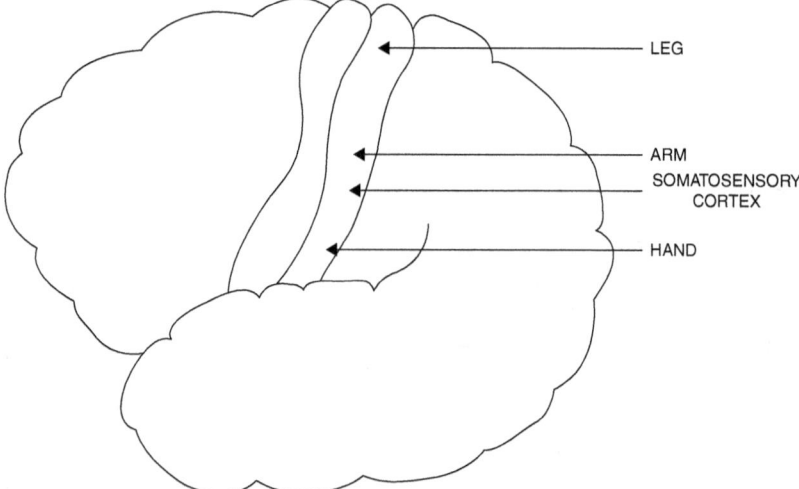

Figure 1.3 **Somatosensory cortex.** The somatosensory cortex is directly behind the central sulcus that separates the frontal and parietal lobes. This area is also called the postcentral gyrus. It receives sensory information from the skin (e.g., touch, temperature) and joints (position sense). Injury to this area impairs patients' ability to recognize objects that they touch as well as the direction and magnitude of joint movements. The lowest part of the postcentral gyrus obtains information from the face and hand, and the highest part obtains information from the lower extremity. This information is transmitted through a relay station deep in each hemisphere called the thalamus.

Introduction 5

of visual stimuli. The somatosensory cortex performs a similar analysis for stimuli that touch the skin as well as joints and muscle movements. Sensations from the left side of the body go to the right hemisphere's somatosensory cortex, and those from the right side of the body go to the left hemisphere. Sensations from different parts of the body go to different portions of the somatosensory cortex (Figure 1.3), with the face and hand being the lowest and the foot and leg being the highest.

Each of these primary sensory areas in the cerebral cortex (Figure 1.4) sends this analyzed information to their modality-specific sensory association area. The primary visual cortex sends information to two different sensory association networks (Figure 1.5). One is lower down in the brain and is called the ventral (bottom) stream, and the other is higher up and is called the dorsal (top) stream. These association areas perform visual perceptual processing. They receive the elementary sensory information from the primary visual cortex and develop shapes and patterns. In addition, the more ventral stream stores memories of previously viewed objects, such as faces, objects, and words. The right hemisphere ventral stream appears to be more important in the perception of faces and the left in the perception of objects and written words. Thus, Mishkin and Ungerleider (1982) called this visual processing network the "What" stream. When this area is injured by a disease such as a stroke, the patient develops a disorder called "visual agnosia." The word agnosia comes from Greek: a = with

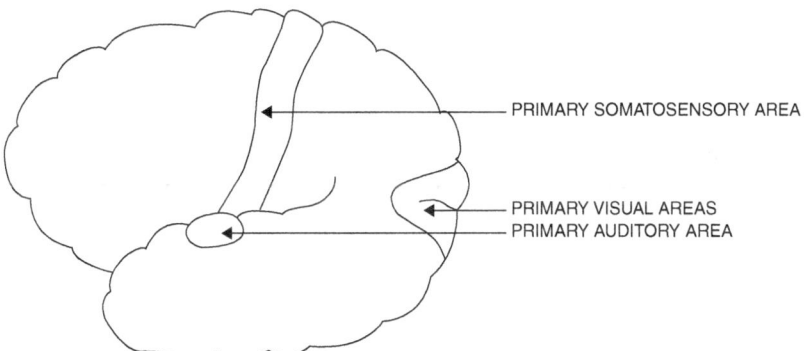

Figure 1.4 **The primary visual, somatosensory, and auditory areas**. Sensory information from the eyes, ears, and skin are all relayed through the thalamus to portions of the cerebral cortex called primary sensory areas. For vision, this area is in the occipital cortex. The tactile information goes to the somatosensory cortex and auditory information to the top of the temporal lobe in an area called Heschl's gyrus. These primary sensory areas primarily perform an analysis of incoming sensory information. For example, when listening to a person speak, Heschl's gyrus will analyze the amplitude, duration, and frequency of incoming information but not recognize the meaning of sequence of phonemes that compose words.

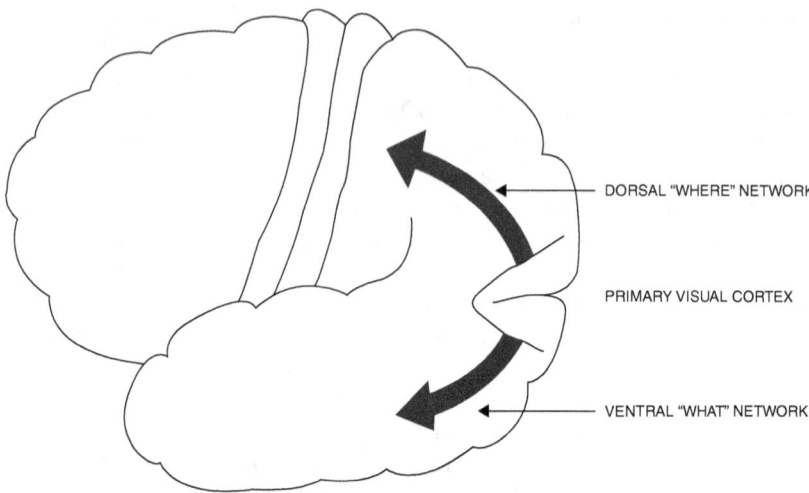

Figure 1.5 **Dorsal "where" and ventral "what" visual networks.** After visual information from the eyes comes back to the occipital cortex and is processed by this cortex, information is sent to visual associations areas. In general, there appears to be two major visual processing networks. One of these networks that is higher (dorsal) appears to be important in determining spatial location (i.e. where, in relation to the viewer's body, a visual stimulus is located). This network has been called the "Where" system. The other lower (ventral) network is important for being able to recognize objects (including faces) and has been called the "What" network.

and gnosis = knowledge. Patients who have visual object agnosia can see and have good acuity, but with object agnosia, they cannot recognize objects (Lissauer, 1890). Patients with prosopagnosia cannot recognize the faces of people who they know. Patients with these injuries are impaired in recognizing objects or faces because the memories of how objects look are stored in these regions. However, if a person with prosopagnosia hears the voice of someone they know, they will be able to recognize this person.

Bálint (1909) described patients who could recognize objects, but when attempting to point to these objects or pick them up with their hand, they did not appear to know their location. He called this disorder "optic ataxia." He also reported that these patients had trouble moving their eyes to the correct position to see an object. Typically, patients with this disorder have lesions on their dorsal (top) parietal and occipital lobes. Mishkin and Ungerleider (1982) called this dorsal processing network the "Where" stream.

These visual, auditory, and somatosensory areas all send connections to the inferior parietal lobe. Therefore, this area is called the polymodal cortex. In the inferior parietal lobes, there are interactions between these sensory networks that allow people to know how to successfully interact with their environment. The left hemisphere's inferior parietal lobe is also more important in the spatial and temporal aspects of movement programming, and the right hemisphere's inferior parietal lobe is important in the allocation of attention. Both inferior parietal lobes are also critical for many cognitive activities.

Motor and Premotor Cortex

Motor Cortex: The frontal and parietal lobes are separated by a large fissure called the central sulcus (Figure 1.6). At the very back of the frontal cortex is the primary motor cortex (Figure 1.6). The primary motor cortex contains the nerve cells, which have long axons that travel down through the cerebral hemispheres and the brain stem. These axons then travel in the spinal cord until they connect with motor neurons in the spinal cord. This large cable is called the corticospinal track, and an injury to this track anywhere along its pathway causes paralysis. This motor cortex starts just above the temporal lobe and goes to the top of the hemisphere and partly

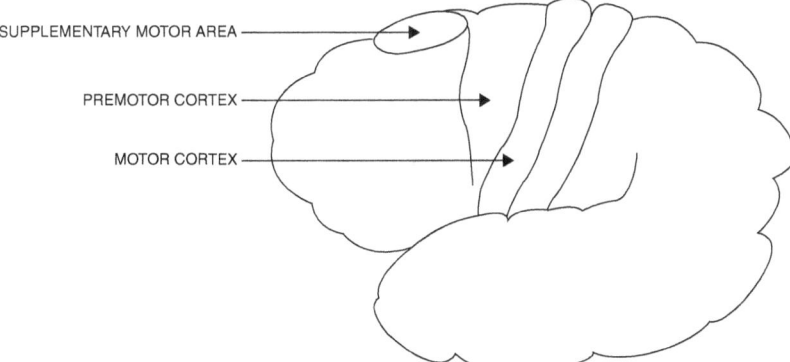

Figure 1.6 **Motor and premotor cortices, including convexity and supplementary motor area.** The primary motor cortex is just in front of the central sulcus. It contains large neurons whose axons travel to the brain stem and excite the motor cranial nerves that control the movements of structures such as the face and tongue. Some of the axons also go to the spinal cord and activate the motor neurons responsible for moving structures such as the arms and legs. The group of axons from the motor cortex down to the spinal cord is called the corticospinal tract. The motor neurons are excited by neurons in the premotor cortex that are just in front of the primary motor cortex. There are two sections of this premotor cortex: one on the lateral convexity of the frontal cortex, just in front of the primary motor cortex, and the second is the supplementary area that is higher up and in the midline portion of the frontal lobes. The premotor cortex also projects to motor neurons in the spinal cord, but for the most part, these control the muscles in the trunk. The premotor cortex also sends projects to the primary motor cortex, and these projections are thought to be important in programming the activation of the neurons in the primary motor cortex.

down the middle (Figure 1.6). This motor cortex is anatomically arranged so that the neurons in the bottom part of the motor cortex go to the motor neurons in the brain stem that control the face, and above these neurons are those that go to neurons in the spinal cord that activate the muscles in the hand and arm. On the top of this motor strip are the neurons that control the leg and foot. The spatial order of the motor neurons in the motor cortex mirrors the somatosensory neurons that are behind them, such that the hand and face are lowest, and the leg and foot are highest up.

Premotor Cortex: The premotor cortex is also in the frontal lobe and is situated just in front of the motor cortex (Figure 1.6). The premotor cortex is the major area of the brain that connects to the motor cortex, and it is responsible for activating the neurons in the motor cortex. Whereas the motor cortex activates the motor nerves in the brain stem and spinal cord, the premotor neurons activate the motor cortex. These premotor neurons activate different parts of the motor cortex either simultaneously or in sequence, allowing us to make the skilled movement of our upper and lower limbs. These premotor neurons, however, do receive projections from parts of the parietal lobe that give them instructions for making the temporal and spatial patterns of movement as well as from areas of the sensory cortex.

Prefrontal Cortex

The portion of the brain that is in front (anterior) of the premotor cortex is called the "prefrontal cortex." This portion of the human brain does not receive any direct sensory input and has no direct motor output, but it has many connections with other portions of the brain, such as the sensory association areas, the polymodal areas, the premotor areas, and the portions of the brain called the limbic system that produce and control emotions as well as memory. Many neuroscientists and clinicians call the prefrontal areas the executive center because it is responsible for planning and initiating activities. Many of the functions of the frontal lobes will be described in the chapters that follow.

Subcortical Areas

Many of the axons coming from the cerebral cortex travel below the cerebral cortex. Like electrical wires, these axons have insulation called myelin, and these myelinated axons below the cortex make the subcortical area look paler than the cortex. Therefore, many neuroscientists and clinicians call these subcortical regions the "white matter" and call the cerebral cortex the "gray matter." There are also many axons that connect the right hemisphere to the left hemisphere and vice versa. This large communication channel is called the corpus callosum (Figure 1.7).

Whereas most of the neurons in the human brain are in the cerebral cortex, there are groups of neurons deep in each hemisphere that form

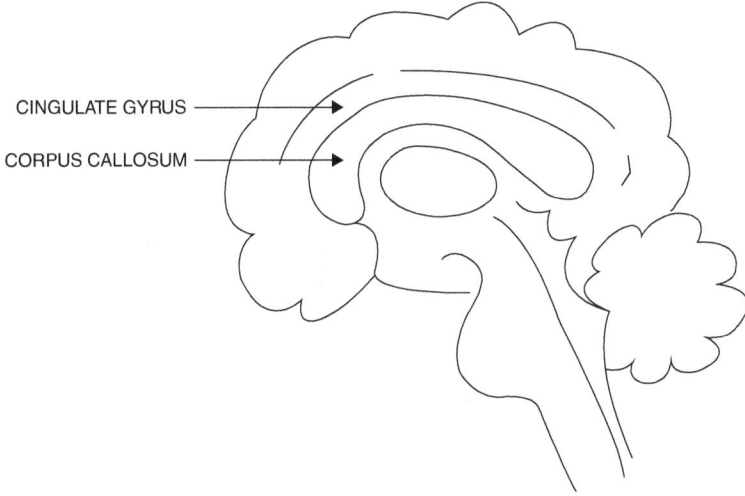

Figure 1.7 **Corpus callosum**. In this figure, the brain is cut in half from front to back. The figure shows the middle of one hemisphere. The corpus callosum is the primary cable that connects the right and left hemispheres. Most of the axons that compose this cable are from neurons that are situated in the cerebral cortex and travel as well as connect with the neurons in the cerebral cortex of the opposite hemisphere. This corpus callosum is critical for interhemispheric communication. This communication can be informative, excitatory, or inhibitory.

a structure called the thalamus (Figure 1.8). In addition to other functions, the thalamus is primarily a relay station for information being transmitted to the cerebral cortex. There are other groups of neurons in structures called basal ganglia (Figure 1.8), which receive information from the cortex and after processing this information send it back to the cortex through the thalamus. While the role of the basal ganglia is not entirely understood, they appear to be important in the control and regulation of movements. Diseases of these basal ganglia, such as Parkinson's disease, appear to cause movement disorders, with impairments in the control of voluntary movements and the occurrence of involuntary movements.

Cerebellum

Under the back (posterior) portion of the cerebral hemispheres is a portion of the brain called the cerebellum, which in Latin means "small brain" (Figure 1.9). The cerebellum, like the cerebrum, has two hemispheres: one on the right and the other on the left. Joining the two hemispheres of the cerebellum is a structure called the vermis, and at

10   *Introduction*

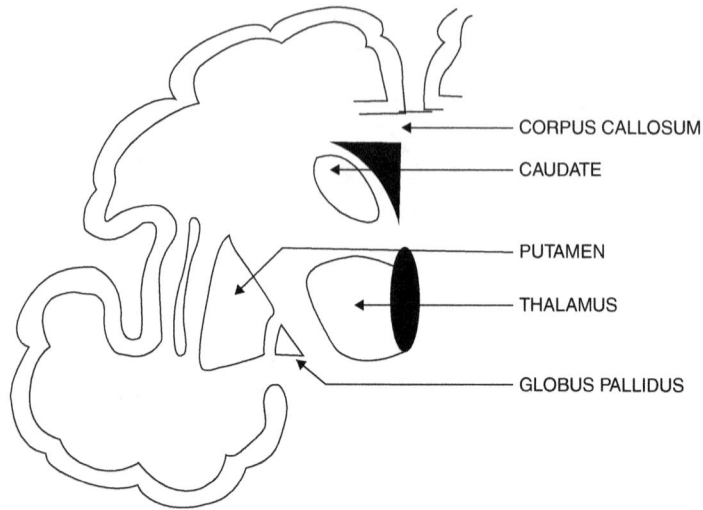

Figure 1.8 Cartoon of the structures deep in each hemisphere, including the basal ganglia (caudate, putamen, and globus pallidus) as well as the thalamus and corpus callosum.

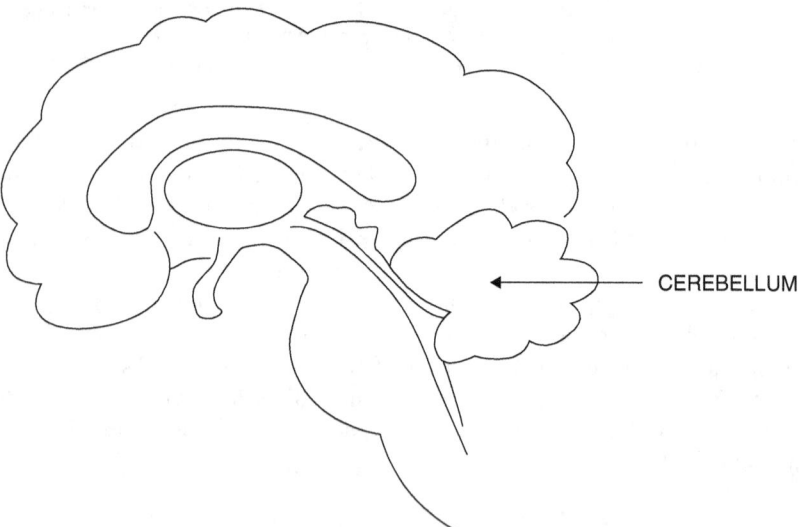

Figure 1.9 The cerebellum.

the bottom of the vermis is another lobe called the flocculonodular lobe. The flocculonodular lobe appears to be important in balance and spatial orientation. Each of the cerebellar hemispheres is divided by the primary fissure into anterior and posterior portions. The anterior portions appear to be important in motor control. The parts that are closer to the middle are important in the control of the body's trunk, and those more lateral are important for the control of the limbs.

Patients with an injury to a cerebellar hemisphere primarily make errors in the magnitude and amplitude of their movements as well as the speed and force of their movements. For example, patients with an injury to the anterior region on one side will have problems accurately moving their arm or leg on the same side of their body as the cerebellar injury. They will often overshoot the target, a phenomenon called dysmetria, and after making a spatial error, their attempts to correct this movement may lead to additional errors, a phenomenon called ataxia. These patients may also have problems making rapid alternating movements of their arm or leg on the same side as the cerebellar injury.

Although not fully explored, the posterior portion of the cerebellum has extensive connections with the cerebral cortex and appears to play a role in mediating certain aspects of speech, spatial attention, and emotions.

Types of Sports

There are a multitude of different types of sports; however, like animals or foods, some sports share similar characteristics with other sports, and therefore the skills needed for these sports may be in some ways similar.

One of the types of sports is *balance sports*. In these sports one of the major required skills is to maintain balance and adjust the body's relationship to gravity. Some sports that are heavily dependent on balance skills include surfing; skiing; diving; mountain bike riding; and several gymnastic events, such as the balance beam.

Another type of sports can be classified as *speed sports*. These include running, swimming, rowing, ice-skating, and cross-country skiing.

A third type of sports is *accuracy sports*. These include bowling, archery, shooting (e.g., rifles), golf, billiards, and table pool.

Combat supports include the different forms of wrestling, boxing, fencing, judo, and kickboxing.

Whereas almost all sports require strength, *strength sports*, such as weight lifting, focus on the athletes' strength.

There are also many types of *ball sports*, and these sports can be subdivided. There are those that use racquets, such as tennis, ping-pong (table tennis), racquetball, badminton, and squash. There are those that use a bat and ball, such as baseball, cricket, and stickball. There are those that require target accuracy, such as soccer, basketball, hockey, and lacrosse, and there are those that use a net that the ball must go over, such as tennis and table tennis, volleyball, and badminton. Finally, there are ball sports that require the capture of territory, such football and rugby.

Almost all sports are not fully physical but require cognitive skills that allow the athlete to develop strategies that can enhance his competitive performance.

In this book, we will describe many of the brain mechanisms that are important for the skills mentioned in this chapter.

2 Movement Action Programming

All sports depend on the athlete's movements. All movements of the human body are produced when a muscle or muscles either contract or relax. The contraction and relaxation of muscles are primarily controlled by the motor nerves that originate in the spinal cord; like a wire, these motor nerves carry messages from the spinal cord to the muscle. These neurons in the spinal cord fire when they get a message from the brain to fire. As mentioned in the introduction, the messages from the brain also come from neurons that have their cell bodies in the brain. These cerebral motor neurons are located at the back of the frontal lobes in an area called the motor cortex (Figure 1.5). These motor neurons send long arms called axons, which act like wires, down through the cerebral hemisphere, down though the brain stem, and into the spinal cord. For movement of the feet, they go way down to the bottom of the spinal cord, but for movements of the arm, they go to the top of the spinal cord. This cable of wires is called the corticospinal tract. The lowest portion of the brain is called the medulla. These cables cross over to the other side, in the medulla and then continue down to the spinal cord to meet the motor nerves that send their messages to the muscles (Figure 2.1). Since this corticospinal cable (track) is crossed, for the most part, the motor cortex on the left side of the brain controls the muscles on the right side of the body, and the motor cortex on the right side of the brain controls the muscles on the left side of the body.

The Praxis "How" System

There are two major types of brain programs that govern our actions and allow us to achieve our athletic goals. One type of programs is responsible for programming "how" to move, and the other is responsible for programming "when" to move. Neuroscientists call the networks that control these "how" programs the "praxis system." For example, the brain contains the programs that allow a person to know "how" to position or posture their hand, arm, and leg when performing athletic activities. In addition, these praxis programs allow people to know "how" to move their joints when performing an athletic activity, such as shooting a basketball or hitting a baseball.

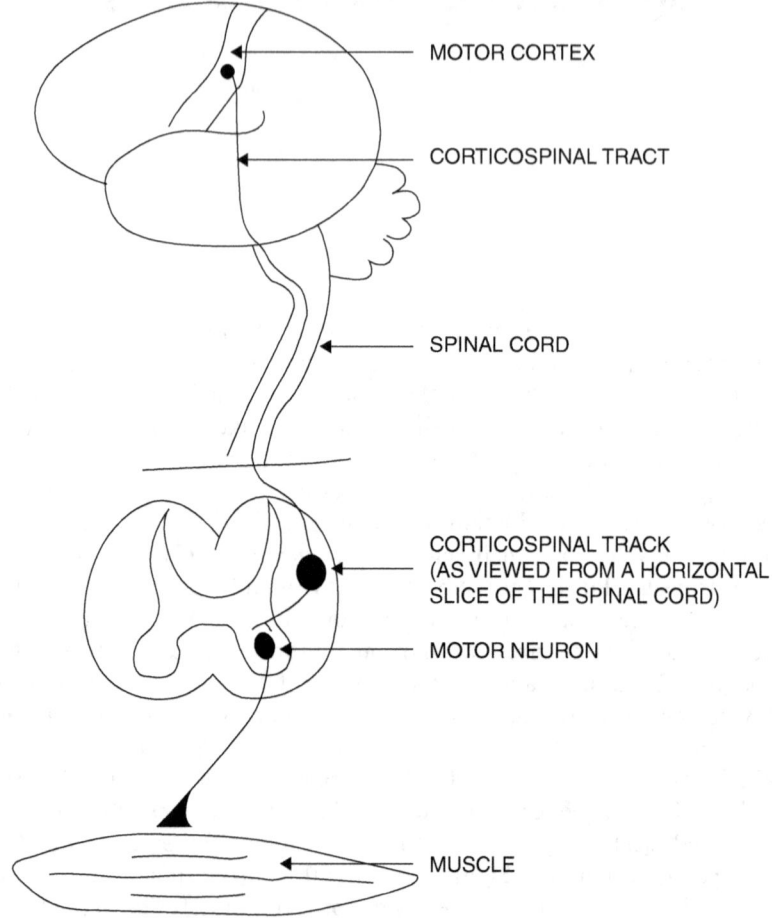

Figure 2.1 **Motor cortex, corticospinal tract in the brain and in spinal cord, motor neuron in spinal cord with motor nerve synapsing with muscle.** Our joint structure, the muscles that move these joints, and the nerves that activate these muscles permit us to make almost an infinite number of different movements, and these movements can occur at different speeds and with different forces. In order to be successful, athletes need to have movement programs that will allow them to be successful at their selected sport.

Almost all athletic actions require the movements of several joints, and these praxis programs allow the athlete to know "how" to move several joints in relation to each other. These praxis programs also allow us to control "how" fast to move each of the joints as well as "how" much force will be needed to successfully complete the action (Heilman and Rothi, 2012). Therefore, for the athlete to perform the purposeful skilled movements required by the specific sport, certain areas of the brain have to both acquire and use the knowledge of "how" to program these learned skilled movements by activating the correct motor neurons in the motor cortex. In addition, the athlete needs to know "how" to correctly activate the correct sequence or series of motor neurons and deactivate other neurons. For example, in order to throw a dart at a target on the wall, first, the muscles that control the flexors of the thumb and forefinger must contract so that these fingers can hold the dart. Then, the neurons in the motor cortex activate the motor neurons in the portion of the spinal cord in the neck (cervical spinal cord) that send motor neurons down to the muscles in the arm. These activated neurons in the spinal cord then send messages down to the motor nerves, which slowly excite the biceps muscle in the arm, which contracts and slowly bends (flexes) the arm at the elbow. After the arm is bent (flexed) at the elbow, the motor cortex stops sending messages to the neurons in the spinal cord that excite the biceps and activates the neurons that go to the triceps muscles in the arm, making these muscles rapidly contract, which causes a rapid straightening (extension) of the arm at the elbow. A different area of the motor cortex also sends messages down the corticospinal tract to the neurons that activate the muscles that make the wrist bend forward. In addition, at the correct time, the neurons that when firing, caused the fingers to contract and hold the dart stop firing then the muscles that extend these fingers fire and release the dart.

Much of the knowledge about the control of movements comes from studying patients who have injuries to specific parts of the brain caused by diseases such as a stroke. Clinicians call the failure to correctly perform purposeful skilled movement apraxia; however, to diagnose an apraxia, this failure cannot be related to weakness; sensory loss; cognitive-thinking impairments; or involuntary movements, such as a tremor. In order to better understand how the athlete's brain may program the multiple aspects of a skilled movement, I will describe motor programming disorders or apraxia.

Apraxia has been studied for more than 100 years, and it has been learned that it is not a single homogenous disorder; rather, there appears to be a variety of apraxic disorders. These different forms of apraxia or loss of "how" knowledge are perhaps best defined by the types of errors made by the patient and the conditions that elicit these errors.

When throwing a dart, instead of moving the wrist forward as the arm is extending at the elbow, a patient with apraxia may incorrectly rotate his arm at the elbow so that the dart does not fly straight ahead but instead heads off to the left of the thrower. Patients who fail to use the correct joint or who fail to properly coordinate joint movements have a disorder that a famous European neurologist, Hugo Liepmann (1920), called *ideomotor apraxia*. The other errors that patients with ideomotor apraxia can make include using the incorrect posture. For example, the dart thrower may hold the dart in a closed fist rather than between their thumb and their index or middle finger.

Patients with ideomotor apraxia may also make speed-time errors; thus, when throwing the dart, they may not move their arm at the correct speed to allow the dart to reach the dartboard, and instead, the dart may fall to the floor.

As we mentioned in the introduction, the brain has two hemispheres. Each hemisphere controls the movements of the limbs on the opposite side of the body, and a large cable-like structure called the corpus callosum (Figure 1.6) connects the two hemispheres. The corpus callosum allows the right and left hemispheres to communicate. Although Liepmann and Maas (1907) were the first physicians to describe an apraxia in a patient with injury to the corpus callosum, the patients they reported had a prior stroke, and thus, their report was confounded by this prior injury. However, Bob Watson and I (Watson and Heilman, 1983) examined a patient who had a vessel with a weak wall (aneurysm) explode. The blood that was released from this vessel traveled to the region between the two hemispheres, and it caused the vessel that supplies blood to the corpus callosum to go into spasm. This spasm prevented blood from reaching the corpus callosum, and therefore, without blood, this cable was injured and could not transmit messages between the right and left hemispheres.

This woman did not have weakness on either side of her body, and therefore, we could test both of her arms. We asked her to pantomime with her arm and hand several different movements needed to use tools such as scissors, a screwdriver, a hammer, and a knife. She performed all these pantomimes flawlessly with her right arm and hand. In contrast, when we asked her to perform skilled learned movements with her left arm and hand, she performed terribly, holding her hand in incorrect postures and moving the incorrect joints. Her performance was so poor that it was difficult for Bob Watson and me to even decide if she was making the correct movements.

In the mid-nineteenth century, there was a famous French physician, Paul Broca, who studied patients who had language-speech impairments from strokes that injured the cerebral hemisphere. In his first paper on this subject, Paul Broca wrote about a man who lost the ability to express himself with speech but who could comprehend other people. When this man died, Paul Broca (1861) found that the man had damage to the left front part of his brain. Paul Broca also collected a series of patients who were right-handed and who had impaired speech (aphasia) from a stroke. He found that all these patients had damage to their left hemisphere. Based on this report and subsequent studies, it became known that in most humans, it is the left hemisphere that mediates speech.

In 1962, Norman Geschwind, together with Edith Kaplan, reported a man who had a corpus callosum lesion as a complication of surgery. The surgeon was attempting to remove a brain tumor and clamped a vessel that supplies blood to the corpus callosum. When Norman and Edith asked this man to pantomime skilled movements with his right arm, he performed these movements very well; however, when asked by verbal command to pantomime purposeful skilled actions with his left hand, he performed poorly. In contrast, when he was asked to imitate the examiner performing skilled purposeful movement, he performed normally with both hands. In addition, he could properly use actual tools and implements with both hands. Since this patient had a lesion of the connecting cable, the corpus callosum, he was unable to perform correctly with his left arm and hand in response to verbal commands but was able to imitate and use actual tools; Geschwind and Kaplan (1962) thought that his left upper limb apraxia was related to a verbal disconnection. Although his left hemisphere could understand the commands, it could not transfer this information to the right hemisphere that controls the left arm and hand; imitation and the use of actual tools do not require language comprehension, and thus Geschwind and Kaplan's patient was able to perform these actions. Had their patient been an athlete, he probably could have thrown a ball normally with his right and left arm.

The patient that Bob Watson and I described (Watson and Heilman, 1983) was very different from Geschwind and Kaplan's patient. Even when we asked her to imitate our performing a skilled purposeful movement with her left hand, she made many spatial and temporal errors. When we gave her actual tools and implements to use with her left hand, she made many errors. This woman's right hemisphere was intact, and therefore her deficits when using her left arm and hand could not be accounted for by a primary motor or sensory disorder.

When we gave her a hammer and asked her to show how she would pound a nail into a board, she performed very well with her right upper limb. We then put the hammer in her left hand and asked her to show us how she would use this hammer to pound a nail with her left arm and hand. She said to us, "For your safety, you may want to get out of this room." We told her, "Just try it." Her movements were so wild that Bob and I were ready to crawl under her bed, but fortunately, Bob was able to get this hammer out of her hand and ensure our safety. After our study was published, there were subsequent reports by other investigators that confirmed our findings.

Based on their observation of a patient with callosal injury, Liepmann and Maas (1908) suggested that the left hemisphere of right-handed people was not only dominant for mediating speech and language but also contains the neuronal networks (representations) that stored the knowledge needed to program learned skilled purposeful movements. Liepmann (1920) called these representations "movement formulas" and stated that they contain the "time space form picture of the movement" (spatial-temporal movement representations). Our patient, with a hemispheric disconnection from a lesion of her corpus callosum, certainly appeared to support Liepmann's hypothesis. He also thought that this asymmetrical storage of movement representations was the basis of hand preference. However, handedness and athletic abilities will be written about in another chapter.

Further support of this hemispheric laterality of movement representations postulate was provided by Liepmann's (1920) study of a large population of patients who had strokes of their left or right hemisphere. He found that ideomotor apraxia was much more common with left than right hemisphere strokes, even when the arm and hand on the same side as the stroke were tested. Subsequent studies have supported Liepmann's laterality hypothesis such that in people who are right handed, almost all cases of apraxia are associated with left hemisphere lesions (Geschwind, 1965; Goodglass and Kaplan, 1963); however, there are some right-handed people who developed ideomotor apraxia from right hemisphere lesions, and this suggests that there may be other factors that determine hand preference. This will also be mentioned in the chapter about handedness.

Since the left hemisphere is also dominant for mediating language, some people have thought that perhaps the loss of the ability to correctly perform learned skilled purposeful movements (ideomotor apraxia), caused by left hemisphere lesions, might be related to the language deficits (aphasia) caused by these brain injuries. However, Liepmann (1920) and others have noticed that there are patients with left hemisphere strokes who have apraxia but not aphasia and others who have aphasia but not apraxia. Since motor action and language disorders can be disassociated, a language disorder cannot fully explain the loss of the ability to correctly perform purposeful skilled movements.

From the developmental aspect, there are many people who have wonderful language skills who are terrible athletes in those sports that require temporal-spatial skills of the upper extremities, and there are great athletes who have poor language skills.

Liepmann (1920) noticed that many of his patients with ideomotor apraxia appeared to have injury to the left inferior parietal lobe (Figure 2.2), and therefore, he thought that this was the area in the brain that stored the spatial-temporal representations or movement formulas needed to perform purposeful learned skilled movements.

Almost all coaches, prior to coaching, participated in the sport they are now coaching. Therefore, not only did they learn the skilled movements needed for that sport, but also, they are able to recognize when the athletes they are coaching are performing correctly or incorrectly, and if they are performing incorrectly, coaches can teach their players to perform the movements correctly. If the parietal lobes store the temporal-spatial memories of learned skilled movements and if this area is damaged, then not only should this person be impaired at performing

Movement Action Programming 19

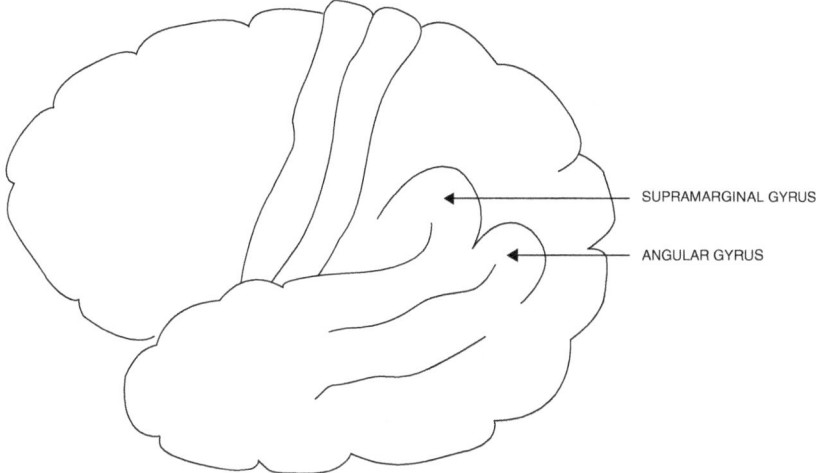

Figure 2.2 **Supramarginal and angular gyri of the inferior parietal lobe.** The supramaginal gyrus, which is located at the end of the ascending branch of the sylvian fissure, and the angular gyrus, which is located at the end of the superior temporal sulcus, receive inputs from the auditory, visual, and somesthetic association areas, and thus these areas have been called polymodal or heteromodal areas. Damage to these areas in the left hemisphere may impair reading, writing, and arithmetic, as well as knowing right from left, the names of the fingers, and how to perform learning skilled movements. Injury to the right inferior parietal lobe can cause unawareness of left hemisphere and the left side of the body called hemispatial neglect.

skilled movements, but they should also be impaired at recognizing other people's errors. We examined patients who had left hemisphere lesions that most likely injured the left parietal lobe versus those with more anterior lesions of the left hemisphere and found that those with the posterior lesions were not only impaired at performing purposeful skilled movements but were also impaired at discriminating between those pantomimed actions that the examiner performed correctly and those that were performed incorrectly.

In the introduction, we mentioned that input from the eye's retina is sent to the back of the brain in the occipital cortex. When we position or move a part of our body, the information about the position and movement is sent by sensory (proprioceptive) receptors that are in our muscles and tendons (Golgi tendon organ) to the spinal cord by sensory nerves. This information is then sent to the brain. The part of the cortex that is in the most anterior (front) of the parietal lobe is the area that receives and processes this information. This sensory receiving area is called the somatosensory cortex (Figure 1.3).

The sensory association areas store sensory memories of previous sensory experiences. For example, if a portion of the visual association area is injured, a person may have good visual acuity but not be able to recognize what they are looking at.

The inferior parietal lobe is called "polymodal cortex" because it receives and synthesizes information from both the somatosensory cortex and the visual association cortex, and when learning to perform purposeful skilled movement, such as that used in a sport, the athlete needs to combine the incoming information from both proprioceptive and visual systems, and store this information.

After visual information comes into the occipital cortex, this information is processed and analyzed by two different visual association areas, and when the ventral (bottom) area is injured, patients have a visual agnosia and are impaired at recognizing objects or faces or both (Lissauer, 1890). The reason patients with these injuries cannot recognize objects is because the memories of how objects look are stored in these regions. Mort Mishkin and Leslie Ungerleider (1982) have called this ventral (bottom of brain) network the "What" system.

Although patients with visual object agnosia cannot recognize the objects placed before them, they have no trouble correctly pointing to or looking at these objects. As also mentioned in the introduction, Bálint (1909) described patients who could recognize objects, but when attempting to point to these objects or pick them up with their hand, they did not appear to know their location; Bálint called this disorder "optic ataxia." The patients with optic ataxia often have lesions to their superior parietal lobe (Figure 2.3), and thus these lesions injured their dorsal (top) visual stream. Mishkin and Ungerleider (1982) call this dorsal stream the "Where" visual system.

Both these visual streams connect with the inferior parietal lobe. Thus, when the athlete, such as a quarterback, throws a pass, his movement formula provides information about the need to: (1) bend his hand and fingers around the football and flex the fingers so that the football is securely held by the hand, (2) bend the wrist so that the thumb first moves toward the body, (3) flex the forearm at the elbow, and (4) lift the arm at the shoulder and then reverse this pattern of movements. In addition to this "how" knowledge, the quarterback needs to identify his receiver and the position of his receiver or where the receiver will be when the ball approaches him. This "where" localization information gives the movement representation about the needed speed and direction of this skilled movement.

In some respects, these movement formulas stored in the left inferior parietal lobe are like sheet music stored in the brain. If you are playing the piano, the sheet music provides the information as to where you should position your fingers (so they can hit the correct keys), the order in which they should press the keys, how long they should remain pressing

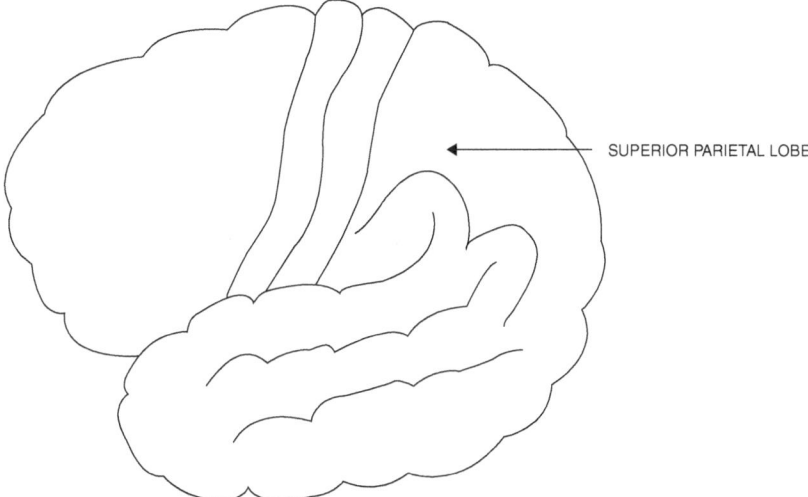

Figure 2.3 **Superior parietal lobe.** The superior parietal lobe is above the supramarginal gyrus and the angular gyrus. It is behind the somatosensory cortex and in front of the occipital cortex. This area appears to be an important part of the dorsal "Where" visual network and contains the somatosensory association cortex, which is important in joint position sense and tactile discrimination.

the keys, and the force that should be used. The posterior (back) part of the frontal lobe contains the primary motor cortex, which in our piano metaphor would be the keys, the connection between the keys together with the hammer would be the motor nerve, and the piano strings would be the muscle. In front of the motor cortex in the frontal lobes is the premotor cortex (Figure 1.6). This premotor cortex is connected to the inferior parietal lobe by a large cable called the superior longitudinal fasciculus (Figure 2.4). The premotor cortex is also connected to the primary motor cortex.

The premotor cortex can be divided into two areas: one that is on the side of each hemisphere and one that is located high up and in the middle of the hemisphere, and premotor areas are just in front of (anterior to) the primary motor cortex (Figure 1.6). The premotor cortex that is higher up and in the middle of the hemisphere is called the supplementary motor area (SMA).

When patients have an injury to the left premotor cortex, they also reveal an ideomotor apraxia such that when attempting to perform a learned purposeful movement, they position and move their joints incorrectly, thereby making spatial and temporal errors (Watson et al., 1986). Although these patients made errors that are very similar to the

22 *Movement Action Programming*

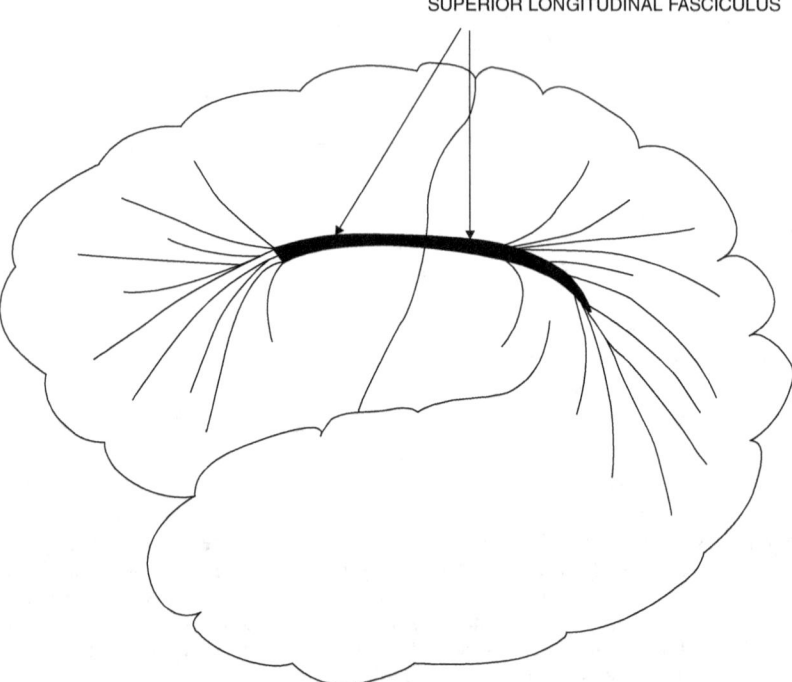

Figure 2.4 **Superior longitudinal fasciculus.** This is one of the major cables that contains the axons of neurons that connect the frontal lobes with the parietal, temporal, and occipital lobes.

patients who had left inferior parietal injuries, these patients, unlike those with parietal lesions, were able to discriminate between correctly performed and incorrectly performed actions. They could recognize these errors because they had intact spatial-temporal movement representations: their sheet music in their left parietal lobe. Thus, normally, these premotor areas appear to be like the pianist who reads the sheet music and, based on this sheet music, directs the fingers to hit the correct keys with the correct sequence, speed, duration, and force of movements.

Over the past several decades, there have been dramatic advances in techniques that allow scientists to learn the portions of the brain that are active during the performance of a task. When a portion of the brain increases its activity, it uses more energy and thus needs more fuel, which is brought to this area by the blood; thus with an increased need for glucose and oxygen, the blood flow to this area increases. When using positron emission tomography (PET) imaging, a radioactive substance is injected into the participants' blood, and a radiation detector is placed around this person's head. These detectors record the changes in radiation emanating from different areas of the head when performing the task and the area that is active gets more blood and gives off more radiation. A more recently developed means of functional brain imaging is called functional magnetic resonance imaging (fMRI). When an area of the brain increases its activity, and more blood goes to that area of brain, there are changes in the amount of oxygen carried by the red blood cells, and this alters the magnetic properties of the blood. The magnetic imaging device detects these changes and demonstrates the areas of the brain that are most active when performing a task, and thus these areas are critical for performing this task. Studies have used functional imaging to learn which area of the brain is the most active when healthy people are performing purposeful learned skilled movements of their upper limbs. When Bohlhalter and coworkers (2009) performed fMRI while healthy participants were performing learned skilled actions with their upper extremities, they found that the left inferior parietal lobe-premotor network described earlier became activated, and this occurred even when the participants were using their left upper limbs. A cartoon of this network is demonstrated in Figure 2.5.

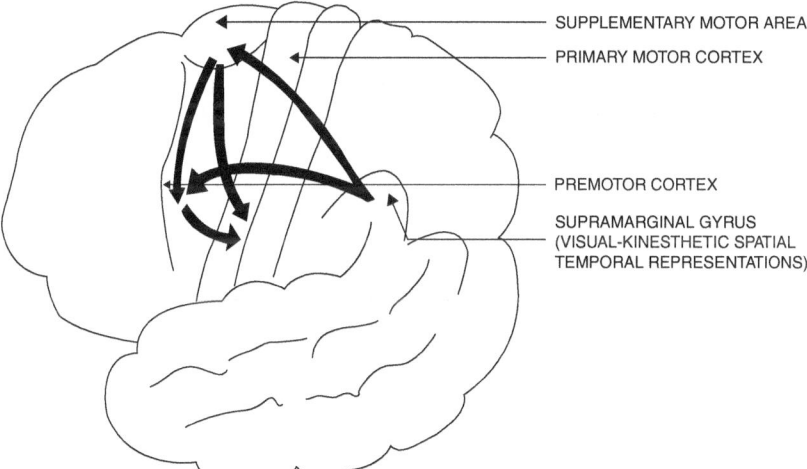

Figure 2.5 **Praxis network.** In right-handed people, the praxis network is important in the programming of learned skilled purposeful movements. The supramarginal gyrus in the inferior parietal lobe appears to contain the visual and kinesthetic spatial and temporal representations (memories) of learned skilled movements. This area is connected by means of the superior longitudinal fasciculus to the premotor areas, both the supplementary motor area and the convexity premotor cortex. These premotor areas convert these visual and kinesthetic representations into motor programs and then selectively activate portions of the motor cortex that are important for activating the motor neurons and muscles needed to make these skilled movements.

Learning New Athletic Motor Skills

Although different people are born with different brains and thus have different potential for developing their athletic abilities, almost all sports require that the athlete learns the actions that are required by that sport.

Motor Memory

The human brain stores information, and this storage is called memory; however, there are different forms of information, and there appear to be different brain networks that store and retrieve these memories. For example, one of the most famous reports about memory loss is the story of HM. HM had epilepsy that could not be controlled with medications. Therefore, his physician thought that if they removed the parts of the brain where these seizures started, this might lead to a reduction in the frequency of his seizures. Since his seizures appeared to be starting from both the left and right anterior (forward) portions of his temporal lobes, the surgeons removed both the right and left anterior temporal lobes. This surgery was successful in that it did reduce the frequency of his seizures; however, this removal caused a terrible disability. He lost his ability to form new episodic memories (Scoville and Milner, 57). For example, he could not recall what he ate the night before or the person with whom he had dinner. Subsequent studies revealed that the critical area removed by the surgeons was a structure in the middle part and at the bottom of temporal lobes, called the hippocampus (Figure 2.6); however,

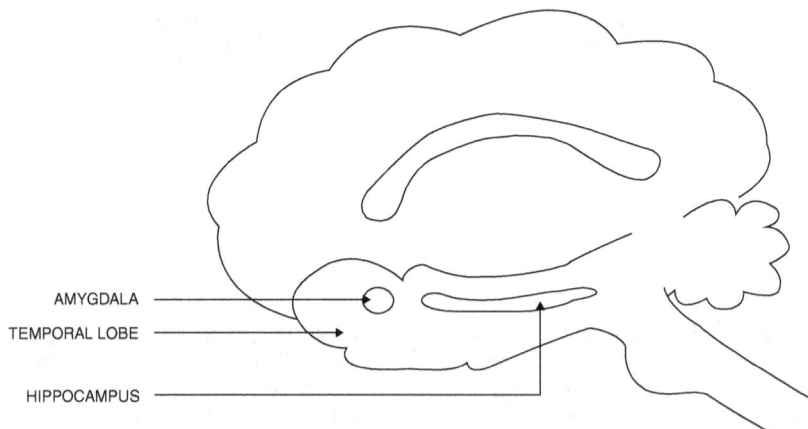

Figure 2.6 **Hippocampus.** This figure is what the brain would look like if it was cut in the middle going from front to back. In the bottom part of this figure is the medial (middle) part on the temporal lobes. The hippocampus is critical for storing new episodic memory (e.g. what we had for dinner last night).

all the knowledge he acquired before this surgery remained, and his IQ testing showed no decline.

Suzanne Corkin wanted to learn if HM could learn new motor skills, so she tested his ability to learn motor skills. For example, one of the tests she performed was called rotary pursuit. In this test, there is a spinning disk, like a record on a turntable. On this disk, there is a small metal target. The person performing this task is given a wand, and their task is to keep the tip of this wand on the metal target, as long as possible, while it is rotating. Typically, when healthy people perform this task each day over several days, they improve this skill, and they are able to stay on target for longer periods. When they tested HM on this task, the same investigators had to reintroduce themselves each day because he could not recall ever meeting them. In addition, they had to repeatedly explain the task; however, HM's performance improved with his daily practice. The dissociation of his episodic memory and his ability to learn a new motor skill revealed that learning motor skill uses a different part of the brain than does storing episodic memories.

Further support for this hypothesis came from a second study. Yamashita (1993) studied three patients who were amnesic due to encephalitis and six normal control subjects by having them perform a 45-rpm rotary pursuit task. Bilateral damage of the medial temporal lobe was confirmed by MRI for all three patients. All amnesic patients acquired the skill. Even on the retention test, after a seven-day interval, amnesic patients showed complete retention of the skill without recalling their training.

Many years ago, we did a study to learn if in right-handed people, it is the left hemisphere that is important in learning motor skills. Each hemisphere controls the opposite hand. We had healthy right-handed participants attempt to learn a series of meaningless movements with their right or left hand, and then we tested the other hand. We found that when the left hand learned a skilled series of movements, many of the participants performed this learned series of movements better with the right hand than with the left hand. However, when they learned these movements with their right hand, the left hand did not perform as well as the right hand. The results of this study provided support that in people who are right handed, it is primarily their left hemisphere that learns and stores information about motor skills.

Maria Wyke (1971) studied patients with either right or left hemisphere lesions and gave these patients a motor acquisition task that required bimanual coordination. Although she found that the patients with left hemisphere injury demonstrated the acquisition of this skill, their performance was below the level of skill demonstrated by the patients who had a right hemisphere disease. In a prior section, the syndrome of ideomotor apraxia was described. Patients with this disorder can no longer perform the skilled movements that they had learned before their neurological disease. Wyke did not separate the subjects with left hemisphere injury into those with and without an ideomotor apraxia. This would have allowed Wyke to learn if these patients' impairment was related to damage of portions of the left hemisphere that program skilled movements. We (Heilman et al., 1975) studied right-handed patients with both an ideomotor apraxia and aphasia and compared these patients to control participants who had an aphasia but did not have an ideomotor apraxia. Similar to Susan Corkin's study of HM, we gave these patients six trials on a rotary pursuit apparatus to learn if they could learn a new motor skill. These patients had a left hemisphere stroke, and since many had a right hemiparesis, all these patients used their left, non-paretic forelimb to perform this task. We found that the patients without ideomotor apraxia, the control group, performance on the last trial was significantly better than on their first trial. In contrast, the group of patients who had ideomotor apraxia revealed no significant difference between the first and sixth trials. These apraxic patients' motor acquisition defect appeared to be caused by a combined impairment of both skill acquisition and retention. It was thought that deficit was caused by the injury of the region of the brain where normally this information is stored, such as the left inferior parietal lobe. Leslie Gonzalez-Rothi and I thought that if they could no longer perform skilled movement because they damaged the area of the brain that stores these movement memories, they should also have trouble with learning new gestures (Rothi and Heilman, 1984). We studied a series of patients who had apraxia from a lesion in the left parietal lobe to learn if they would have difficulty learning new gestures and, if they had performance deficits, whether these deficits resulted from an inability either to consolidate this information into their stores of procedure-movement memory or to retrieve the information once stored. We found that these apraxic patients did have difficulty acquiring new gestures. This inability to reproduce gestural information was not associated with a retrieval disorder, but instead these apraxic patients could not consolidate this movement information into their memory.

In many sports, people use imagery to help their performance. If patients with ideomotor apraxia have a loss of movement representations, it might be expected that these patients may also demonstrate imagery deficits of previously learned movements. Ochipa et al. (1997) studied patients with ideomotor apraxia and found that this patient had movement imagery deficits. In contrast, this patient's object imagery was intact.

When I was about 50 years old, I decided that I was going to learn to play golf. I liked the idea that this is a sport you can play by yourself and that you compete against yourself. If you walk around a course, you get some nice exercise, and many of the courses where I live are beautiful. In general, when I played other sports, such as tennis, I was never an outstanding athletic, but my playing ability was respectable. Before I started to play golf, I said to myself, "How difficult can this sport be? After all, yes, you have to hit a small ball, but unlike other sports, the ball is standing still." However, even after lesions and hours of practice on driving ranges and golf courses, as well as my playing this sport about once a week for more than 25 years, I am still a terrible golfer. It is unusual for me to break a score of 100 (28 strokes over par), and when I do break 100, I celebrate. In all the years I have played, I only scored below 90 once and saved this scorecard. As a neurologist and researcher, I want to know why?

One of the first investigators to study age and motor learning was William Thorpe (1958). He studied young bird's ability to learn their songs. Similar to human speech development, some birds learn to sing using a two-stage process. Songbirds first form an auditory memory from hearing adult birds sing their song. Then, these young birds learn how to program their motor system to produce the sounds stored in their memories. Thorpe found that chaffinches who were raised in a laboratory and never exposed to adult male chaffinches singing their songs never learned their songs; however, if recordings of adult chaffinches singing were played while they were young, they could learn these songs.

Also, like humans, some songbirds can have a regional dialect. Peter Marler and his coworkers (1970) showed that the song dialect is also learned during these birds' sensitive learning period. West and King (1988) reported that learning these songs and the dialect enhances their mating behaviors.

As mentioned in the previous chapter, in the brains of all animals, there are nerve cells, also called neurons. Neurons interact with other neurons through their branches. Like a tree, there are two forms of branches: short branches that are called dendrites and long branches that are called axons. Neurons communicate with other neurons by giving off chemicals called neurotransmitters. Some neurotransmitters given off by the end of neurons increase the activation of the neurons to which they are connected. In contrast, other neurotransmitters can inhibit (decrease) the activity of the neurons to which they attach. Memories are stored in the brain by the patterns of connectivity between neurons. Donald Hebb (1949), the famous Canadian neuropsychologist, stated that "neuron that fire together wire together." As these songbirds mature, the programs that produce these songs become crystallized, meaning that the neuron networks storing the memories of the sounds of the songs and the means to produce these songs cannot be altered. Research has revealed that as these birds mature, they increase their production of testosterone, and it is the hormone testosterone that reduces their ability to alter the networks that are responsible for programming their songs. If birds are given testosterone at an early age, they may have trouble learning to produce their song. If birds are castrated and thus do not increase their testosterone, they can even learn their song at a later age (Bottjer and Johnson, 1997).

What about humans? When Henry Kissinger was Secretary of State, I noticed that while he spoke good English, he had a strong German accent. He was born in Germany in 1928, and he and his family left Germany in 1938 because of the Nazi anti-Semitic programs. When he came to America, he still was young enough to attend high school in New York City, but he had likely already gone through puberty, which is associated with an increase of testosterone, and perhaps like the songbirds, this prevented him from learning the motor programs that would have produced a New York accent. There have been several studies that have supported observations that people who moved to a new country, where they spoke a different language before puberty, are much less likely to have a foreign accident than those who have immigrated after puberty.

About sports, Earl Woods began training his son Tiger to play golf at the tender age of ten months. You can watch Tiger Woods at the age of two hitting golf balls on YouTube. He appeared on the Mike Douglas television show in 1980 at the age of two and hit golf balls. Viewers were amazed by his beautiful golf swing. Many great athletes, but not all, started playing their sports before puberty. While to my knowledge, there have been no studies that have examined the difference in athletic skills for any specific sport comparing those who learned this sport before puberty to those who learned this sport after puberty, based on the studies of birds and foreign accents, I would suspect that those who learn a sport such as golf before puberty would be more likely to have lower scores than those who learned after puberty. Arnold Palmer began playing golf at age four. Ben Hogan was nine years old, and Jack Nicklaus started playing golf before he was ten.

As I mentioned, I started playing golf at the age of 50. If I had tried to learn after puberty, when I was in my teens or twenties, would I have learned to play golf better than when I was in my fifties? I think the answer would be yes, but why?

As mentioned, the patient HM, who had epilepsy and had had his temporal lobes removed, including the hippocampi, had a severe episodic memory deficit. HM was, however, able to learn new motor skills. When I found that I was having difficulty learning to play golf well enough to break 100, as a neurologist, I wanted to learn what may be happening with my aging brain that may be impairing my ability to learn this sport. As a neurologist, I looked at what diseases may interfere with the learning of new motor skills and learned that patients with Parkinson's disease often have problems learning new motor skills.

Patients with Parkinson's disease have a loss of the cells that make the neurotransmitter dopamine. The human brain has a motor cortex that sends messages to the spinal cord, and activation of the neurons in the spinal cord sends messages by means of the motor nerves that activate the muscles that make the limbs and body move. The motor cortex receives programming information from the premotor cortex. The motor and premotor cortices make a circuit and send messages to the basal ganglia, groups of neurons that are below the cerebral cortex. Then, these basal ganglia send messages to the thalamus, another group of neurons that are deep in the middle of each hemisphere. Finally, the thalamus sends information back to cerebral cortex. The neurotransmitter dopamine that is manufactured in the midbrain is sent to these basal ganglia, and patients with Parkinson's disease who do not manufacture sufficient dopamine have movement disorders. While it is not entirely clear what the function of the basal ganglia may be, patients with Parkinson's disease have trouble spontaneously initiating movements; manifest slowing of their movements; make smaller movements than needed; and cannot properly relax their muscles, so they are still stiff.

Before a person develops the signs and symptoms of Parkinson's disease, that person has to lose about 70–80 percent of the cells in their midbrain that make the dopamine needed for proper function of the basal ganglia motor circuits. However, with aging, there is a loss of these dopaminergic neurons. Since patients with Parkinson's disease have trouble learning new motor skills and have a reduction in their dopamine, I thought that perhaps with aging and a decrease in dopamine that is not severe enough to cause Parkinson's disease, this decrease may still impair learning.

When I was in training, I had patients with severe Parkinson's disease who had such severe problems in initiating movement that they could not even get out of their bed to go to the bathroom. The neurons that make dopamine take a chemical called levodopa and change this chemical to dopamine. In 1968, G. Cotzias published a paper in the *New England Journal of Medicine* and reported that when patients with Parkinson's disease were given levodopa, they dramatically improved. When I gave this medication to a patient who was unable to get out of bed by himself, after taking this medication this patient's activity pattern appeared almost normal. Even though I did not and do not have Parkinson's disease, since I was having trouble learning the motor skills required for golf, if I took dopamine could I better learn to play golf?

I also thought that since this medicine at low doses has minimal side effects, it may be worthwhile learning if older people would improve their motor learning skills by taking this medication. Studies such as this are fairly expensive to perform, and therefore I applied for a research grant. I was very surprised to learn that this study would be funded; however, at about the same time I learned about funding, a paper was published revealing that treating older people with levodopa can help them learn new motor skills. Since the experiment I was planning was already published, I never initiated these studies. There is an old saying in medicine, "Do what I tell you to do, but not what I do." It has been many years since this paper was published, but I still have not taken golf lessons after taking a dose of levodopa, and I am still trying to break a score of 100. However, I have learned that, when learning a new sport, younger is better and have encouraged all my grandchildren to take golf lessons when they were young.

Practice

Does practice make perfect, and what is the role of practice in athletic development? According to Oldenziel et al. (2004), the average numbers of hours needed to develop into world leading athletes in basketball, field hockey, and wrestling are, respectively, about 4,000, 4,000, and 6,000 hours. In addition, it took about 7.5 years for the majority of novice athletes to develop into senior elite athletes; however, there are many outstanding athletes who, within a short time of taking up a new sport, develop into outstanding athletes.

Specialization

In physics, there is the law that pressure equals force divided by the area on which the force is applied. Therefore, it is easier to push a sharped pencil through a piece of cardboard than the head of a hammer. Based on this postulate, I would have thought that if a person just focused on one sport, they would be much better at playing this sport than other people who play many sports; however, there are papers suggesting that early specialization in one sport can have significant negative consequences on the development of an athlete over time. For example, studies have shown that an early specialization can lead to an increase in the probability of discontinuing playing this sport (Gould et.al., 1996) as well as limiting the range of motor skills (Wiersma, 2000). In the introduction, we mentioned that there are several forms of sports, such as speed sports, balance sports, accuracy sports, combat sports, and ball sports. Whereas each of these types of sports has different primary skills, for almost all sports, some forms of motor skills are important, and learning many sports allows the athlete to have a greater repertoire of motor skills that can enhance their performance in whatever sport they are playing.

Precision

In order to be a skilled athlete in many sports, such as billiards, archery, or playing darts, it is not strength or speed that determines the athlete's success but rather hand and finger precision. However, in some other sports, in addition to speed, the athlete also needs finger, hand, and arm precision.

Some people call the ability to make precise, independent but coordinated finger movements dexterity; however, one of the definitions of dexterity is right-handedness, and there are many left-handed people who can make precise, independent, and coordinated left-sided finger and hand movements. Therefore, rather than dexterity, we use the term deftness.

Humans' deftness is remarkable. During the time I was a resident at the Harvard Neurological Unit, I was walking in the hallway with Dr. Normal Geschwind, who was the head of our Department of Neurology. Walking in the opposite direction was a neurosurgeon who looked like he just completed performing a surgical operation. We both knew this surgeon and knew that he did not ordinarily use eyeglasses. However, Dr. Geschwind noticed that he had a pair of goggle-like glasses on his forehead. Geschwind stopped this surgeon and asked him, "What's on your forehead?" The surgeon reached up and said, "Oh I should have taken them off in the operating room." Geschwind again asked, "Why do you use them?" The surgeon said, "Oh they just are magnifying glasses so we can see what we are doing." The surgeon as well as Geschwind and I continued our walking. Geschwind then turned to me and said, "Ken, isn't it remarkable that surgeons are able to make such precise movements that they need magnifying glasses to see their movements."

In the clinic, we often see patients with a variety of neurological diseases who have a loss of the ability to perform deft movements. Hugo Liepmann (1920) called the loss of the ability to make deft movements limb-kinetic apraxia. To test patient's deftness in the clinic, we ask them to perform a simple test. We take a nickel (five cents) out of our pocket and show them how to rotate this nickel between their thumb, index, and middle fingers of each hand as rapidly as possible for 20 revolutions. (Hanna-Pladdy et al., 2002).

Typically, patients with injury to one hemisphere will have a loss of deftness in their contralateral hand; however, these patients often also have weakness of that same upper limb. Therefore, it is difficult to be able to tell if their impaired deftness is related to a motor programming disorder or weakness. Heilman et al. (2000) as well as Hanna-Pladdy and her colleagues (2002) have found that people with right-hand preference who have a left hemispheric injury are more likely to have a loss of deftness in their left hand than right-handed people who have right hemisphere lesions have in their right hand. Hand preference in athletes was discussed in another chapter, but these results do suggest that in right-handed people, the left hemisphere helps to program deft movements for both hands.

Hugo Liepmann (1920) thought that it was injury to the sensorimotor cortex that may induce this disorder. Support for this postulate first came from the neurosurgeon Paul Bucy. Bucy had a patient who had uncontrollable movements of one of her upper limbs. These movements were causing many problems for her, and thus, Bucy thought she might do better if he did surgery that would paralyze that arm.

In the introductory chapter where brain anatomy was reviewed, it was mentioned that in the back part of the frontal lobe, there is an area called the primary motor cortex. This area of the cortex contains neurons that have long branches (axons) that travel down through the hemisphere in a pathway called the posterior limb of the internal capsule. These axons then travel through a large cable called the "cerebral peduncle" into the brain stem. From here, these axons go through the brain stem and then go down the spinal cord until they reach the neurons that are in the spinal cord, called "lower motor neurons." These lower motor neurons go to the muscles in the body and are responsible for sending the signals to the muscles that make them contract. This cable that brings messages from the motor cortex to the spinal cord is called the corticospinal tract. Paul Busy cut this patient's corticospinal tract at the cerebral peduncle. After surgery, this patient had weakness, but her abnormal movements were gone. Eventually, she was able to make movements of this arm but not deft movements of her hand and fingers. This observation suggested that the corticospinal tract is critical for making deft movements.

Additional evidence for the role of the corticospinal tract came from the work of Lawrence and Kuypers (1968), who had Rhesus monkeys retrieve food from holes. Some holes were large, so the monkey could get their entire hand into the hole, and others were so small that these moneys could only get their forefinger and thumb in the hole, and use a pincer grasp to retrieve the food. Before their surgery, they could retrieve food from all the holes, but after their corticospinal tract was cut in the bottom of their brain stem, they could not retrieve small pieces of food from the small holes using the pincer grasp. However, like a young child whose corticospinal system is not fully developed, after several weeks, the monkeys were able to use a palmer grasp where all the fingers are simultaneously flexed against the palm in order to get some food out of the large holes. The monkeys could walk and use their upper limbs to climb because the motor systems that control these functions are not part of the corticospinal tract.

In front of the motor cortex also called Brodmann area 4, there is another area called the premotor area or Brodmann area 6 (Figure 1.5). Other studies have suggested that portions of this convexity premotor cortex are also important for programming deft movements. For example, Fogassi and colleagues (2001) deactivated a portion of the premotor cortex in monkeys, who then had difficulty shaping their hand and fingers to grasp different objects. In addition, Nirkko and coworkers (2001) provided converging evidence for the role of premotor cortex in the programming of deft movements. Using fMRI, they found that discrete unilateral distal finger movements were associated with activation of the contralateral convexity premotor cortex.

As mentioned, right-handed people who have left but not right hemisphere injury lose the ability to make deft movements of their left hand, and the left hand is controlled primarily by their uninjured motor cortex in their right hemisphere. Because the loss of hand finger deftness is on the same side as the brain injury, we call this an ipsilateral limb-kinetic apraxia (Heilman and colleagues, 2000; Hanna-Pladdy and colleagues, 2002). There are least two means by which a hemisphere can control movement of the hand and fingers on the same side. Studies of human's neuroanatomy have revealed that there are asymmetries of the corticospinal system. In addition, most of the fibers in the corticospinal system cross over in the lower part of the brain stem; however, not all fibers cross. Therefore, it is possible that the dominant left hemisphere will have more uncrossed ipsilateral projections than does the right hemisphere. An alternative possibility is that the left hemisphere's motor programming system might influence the right hemisphere's motor system by transmitting information by way of the corpus callosum. Verstichel and Meyrignac (2000) reported a right-handed man who suffered an infarct that injured the anterior and middle parts of his corpus callosum. When tested, this patient demonstrated loss of agility (deftness) of his left hand and these investigators diagnosed that this patient had a "melokinetic" (limb-kinetic) apraxia of his left hand. However, infarctions of the vessel that supplies the corpus callosum also often damage the medial part of the premotor cortex called the SMA. We subsequently examined a patient who had neurosurgery for a cyst that was obstructing the spinal fluid from leaving the ventricles. In order to remove this cyst, the surgeons need to cut through his corpus callosum. Postoperatively, we found that this man had a loss of deftness of his left hand, suggesting that information from his left premotor or motor cortex supplies by means of the corpus callosum important information to the right hemisphere's motor network, and this information is important in the control of deft left-handed movements (Acosta et al., 2014).

Action Sequencing

Athletes must often make a series of movements. For example, when throwing a pass, before throwing the actual pass, the athlete may fake a pass and then turn and throw the pass in a different direction. In the clinic, we use a task developed by A.R. Luria (1966), a famous Russian behavioral neurologist. In the test, both the patient and the examiner are sitting facing each other. The examiner tells the patient, "Please watch what I am doing with my hand and please copy my movements." The examiner who is sitting makes a fist and then hits his fist against the top of his knee, lifts his hand, opens his hand, then straightens the same hand and hits the edge of his hand on the same area above the knee. Then, the examiner again lifts his hand and again turns his hand, so the palm is facing downward toward his knee, then hits his palm again on the same area. The examiner does this thrice and asks the patient to copy these series of movements. The lateral portion of the frontal lobes appears to be important in sequencing and injury to the frontal lobe often impairs performance this test.

Action Recognition

A critical element in many sports is recognizing the actions of both the competitors and even teammates. For example, in football, a defensive back, such as the safety, by watching the quarterback throwing the ball, may be able to tell where it is headed and therefore may decide that rather than follow the split end who is going to the incorrect area of the field, he will go to the area where the ball is likely to land and, by doing so, may be able to intercept the ball.

In neurology, we sometimes see patients who have a condition called agnosia. This term agnosia comes from two stems: "a" means without, and "gnosis" is knowledge. Failures of recognition can be caused by several deficits; however, a patient with agnosia has a failure of recognition that cannot be attributed to deafferentation (a deficit of sensory input to the brain such as blindness, deafness, or a loss of touch). In addition, agnosia is not a naming disorder. For example, when shown a basketball, a patient with naming disorder may be unable to name this object but can show how to shoot this ball. Whereas the patient cannot say "basketball," he still has the knowledge about this object and therefore does not have an agnosia. Agnosia is often modality specific. If a patient with a visual agnosia sees a basketball, he may not be able to know what it is, but if he feels it or sees it bounce, he will recognize that it is a basketball. Visual agnosia can also be domain specific; for example, some patients with strokes that damage the lower back part of the temporal and occipital lobes of their right hemisphere (Figure 2.7), in area of the brain called the fusiform gyrus, may have a disorder called prosopagnosia. People with this disorder cannot recognize the faces of people who they know and who they could easily recognize before their brain injury. They might not even be able to recognize family members. In contrast, when these people hear the voice of someone they know, they might be able to rapidly recognize this person. In addition, these patients can often visually recognize objects. Other patients with injury to the left hemisphere in the same area may not be able to recognize objects but might still be able to recognize faces.

Patients with damage to an area of the left hemisphere called the supramarginal gyrus, which is part of the parietal lobe (Figure 2.2), as mentioned earlier, have a loss of the ability to correctly perform learned skilled movements, a disorder called ideomotor apraxia. It is thought that this part of the brain stores the knowledge of how to perform learned skilled movements (movement representations). Just as we have memories of what objects and faces look like, we also have visual images of movements we have learned and kinesthetic memories (feelings of movements). When this area is damaged, patients are impaired in performing skilled learned movements as well as in imitating these learned movements and even imitating novel meaningless movements. They are also

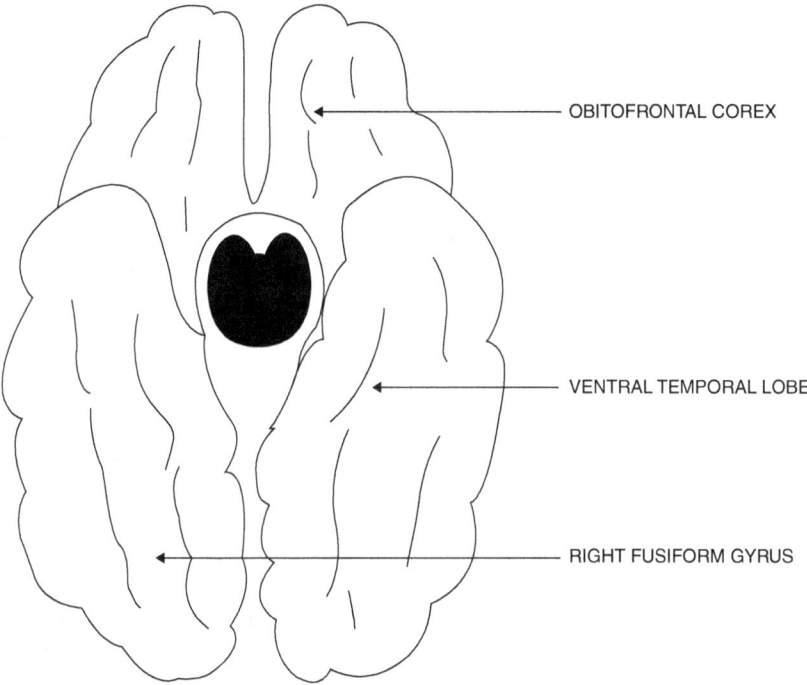

Figure 2.7 **Right fusiform gyrus.** This is a figure of the fusiform gyrus that is located on the bottom (ventral) portion of the occipital lobe and posterior temporal lobe. This visual association area in the right hemisphere stores the visual memories of previously seen faces.

impaired at recognizing well-performed movements as well as discriminating correctly from incorrectly performed movements. For example, I ask some patients if I am correctly demonstrating how to cut a slice of bread from a bread loaf, and instead of making slicing movements, I just make chopping movements, moving my arm only at the elbow. Many patients who have damaged their left supramarginal gyrus will say that I am correctly slicing bread.

Impaired gesture-pantomime discrimination (correct versus incorrect) and recognition that are associated with ideomotor apraxia from lesions of the supramarginal gyrus may not be considered to be a form of agnosia because of the associated production deficits. Based on this strong relationship between performance and recognition discrimination, it may be expected that athletes who have excellent performance will also have terrific recognition. However, there are patients with premotor injuries who are impaired at performance but can recognize the meaning of movements and detect errors. This dissociation would suggest that some people may be excellent at recognizing an athlete's movement but can be poor performers. In addition, Rothi et al. (1986) reported several patients who could not comprehend or discriminate visually presented gestures but who performed gestures normally to command. These patients were able to visually recognize objects. While these patients could be considered to have pantomime agnosia, they could imitate better than they could comprehend or discriminate gestures. Because they could imitate, their inability to discriminate or comprehend gestures could not be accounted for by a defect in vision or visual perception. These patients had left-sided injuries to the back and bottom (ventral) parts of the temporal lobe where it meets the occipital lobe, and these temporal-occipital lesions may have disconnected visual input from accessing the movement representations stored in the parietal lobe. Schwartz et al. (1998) reported patients who could not recognize tools but could recognize pantomimes. These patients were the opposite of those reported by Rothi in that they had object agnosia without pantomime agnosia. This dissociation suggests that the recognition of movements and objects is mediated by different systems and different parts of the brain.

Ungerleider and Mishkin (1982) suggested that there are two processing systems: a ventral "what" system and a "dorsal "where" system. Whereas Rothi et al. (1986) thought that the ventral occipital-temporal "What" visual stream is important for gesture comprehension, the loss of the ability to name tools with a preserved ability to comprehend gestures reported by Schwartz et al. (1998) suggests that it may be the dorsal "Where" visual stream that is important for gesture recognition and the ventral "What" stream that is important for object recognition.

In regard to the neuropsychological mechanism, a model that might help us understand pantomime agnosia comes from the classical aphasia literature. Patients with Wernicke's aphasia and pure word deafness can neither comprehend spoken language nor repeat. Whereas Wernicke's aphasia is thought to be related to the destruction of the phonological lexicon that contains the phonological representations of learned word sounds, pure word deafness is thought to be related to an inability of auditory input to gain access to this lexicon (word-sound dictionary). Patients with transcortical sensory aphasia can, however, repeat (i.e. imitate) in spite of being unable to comprehend words, demonstrating that comprehension and imitation are dissociable. This dissociation between comprehension and imitation suggests that these two processes are at least in part independent and are mediated by different parts of the brain. Lichtheim (1885) suggested that while repetition is mediated by a phonological lexical system that is still functional in patients with transcortical sensory aphasia, comprehension also requires semantic processing. Thus, in transcortical sensory aphasia, the systems that mediate semantic processing of language are impaired or auditory input cannot gain access to these semantic systems (Heilman et al., 1976). The neuropsychological mechanisms underlying impaired gesture comprehension with spared imitation reported by Rothi and Heilman (1985) could be similar. Perhaps these patients can gain access to the input praxicon, but these praxic representations cannot access semantics (Figure 2.5).

Several authors have suggested that speech repetition (i.e. imitation) may be performed either by using stored word representations (lexical) or by using a non-lexical route (just using phoneme, the letter sounds that make up words) (Coslett et al., 1987; McCarthy and Warrington, 1984). Just as we can repeat words we have not heard, we can also mimic movements we have never seen or previously learned to perform. Perhaps imitation, like repetition, can take place without having to access stores of previously learned skilled movements. Support for this alternative imitation system comes from the observation of a patient whose deficit was limited to the imitation of non-familiar limb movements (Mehler, 1987). Since there are no memory stores for unfamiliar movements, the patient had to rely on a nonrepresentational route which was impaired. Some patients with ideomotor apraxia imitate better than they perform to verbal commands, but others do not. Perhaps the patients with ideomotor apraxia who improve with imitation can use this nonrepresentational route, whereas those who do not improve may have an additional deficit in this nonrepresentational route. Some functional imaging research has suggested that gesture imitation may use the same networks that are used with speech imitation and speech repetition (Kühn and Brass, 2008). However, the anatomic differences between these different routes are not known.

The inability to carry out a series of acts, called an "ideational plan," has also been called ideational apraxia (Marcuse, 1904; Pick, 1905). These patients have difficulty sequencing acts in the proper order. For example, when attempting to make a sandwich instead of putting mustard or mayonnaise on one slice of bread, then the ham and cheese, as well as lettuce and tomatoes, followed by putting the second slice of bread on top and then cutting the sandwich in half, the patient with ideational apraxia may first cut each slice of bread in half, before putting on the mustard, ham, and cheese. Most patients with ideational apraxia are suffering with some form of dementia. Both lesion (Mateer, 1999) and functional imaging studies (Wildgruber et al., 1999) suggest that the frontal lobes are critical for sequencing.

Strength

In many forms of athletic events, strength is a critical element. Strength can be divided into two basic forms: endurance and force. Endurance is critical for sports such as marathon running and long-distance cycling. Force is important in sports such as weight lifting and in combat sports, such as wrestling. When athletes train for endurance, they most often use high-repetition but low-resistance exercise over long periods of time. This type of training requires perseverance, and this is described in Chapter 4. Unlike endurance training, strength-force training requires performing exercises that have high resistance (e.g. heavy weights) for short periods of time with a low number of repetitions.

The human body has three types of muscles. One is called smooth muscle, and this is the type of muscle that moves food through our intestines. A second type is cardiac muscle. Although it is critical for athletes to have strong hearts, the type of muscle that allows them to physically move during sport activities is called skeletal muscle.

Skeletal muscles can also be divided into two types. Type 1 muscles are also called slow twitch. These muscles do not create much force but can contract for long periods of time. Type 2 muscles do create much force, but they can only contract for a short period of time. The type of muscle used depends heavily on the type of sport being performed.

Although much can be written about muscles and the motor nerves that activate these muscles, the goal of this book is to review the brain mechanisms of athletic activities, and unfortunately there is only limited knowledge of the role of brain programming on strength. Yes, it is known that for the muscles to contract, the motor cortex in the brain must send down signals to the spinal cord, and the degree of force a muscle exerts is partly dependent on these signals. The means by which these motor neurons in the brain fire determines the strength of contraction of that muscle. Through force-strength training, the brain appears to learn how to best synchronize this firing pattern to obtain the correct force needed. Strength-force training also increases muscle size, and larger muscles can exert greater force; however, studies have revealed that the increase in strength with training is not just based on the growth of muscle. For example, studies have measured the strength of certain muscle groups when performing an action with the upper extremity, before training. Then, the participants in these studies performed strength-force training exercises of this muscle group but only of just one arm. With these exercises, the strength of the muscle group performing this action did increase. However, when the strength in the arm that was not exercising was measured, there was also an increase of strength. Since the muscles in this arm were not exercising, this increase of strength was most likely related to the motor cortex controlling that arm by learning how to best synchronize the firing of the motor neurons that control the muscle being tested. This information learned by one hemisphere may be transmitted to the other hemisphere by means of the corpus callosum, but this hypothesis has not been fully tested.

In addition, to maximize strength, the athlete also has to learn how to relax the muscles that move the limb in the opposite direction as well as to activate those muscles that help support this action.

3 Action-Intention
"When" Programming

Introduction

In the previous chapter, I described "how" we program skilled learned movements, including the spatial, temporal, and force components that are critical components of an athlete's performance. "How" we recognize other people's movements, another critical element of an athlete's performance in many sports, was also described. In order to excel in a sport, in addition to knowing "how" to program skilled movements, athletes need to develop expertise in knowing "when" to act. Whereas humans have an almost infinite number of actions that they can perform, the number of brain processes and actions that can be performed simultaneously is very limited, and the action-intentional system is important in selecting and controlling motor output.

I joined the Air Force during the Vietnam War. Despite being a physician with several years of training in internal medicine at Cornell-Bellevue, I had to undergo basic training. Besides being in Montgomery, Alabama, during the August heat, basic training was fun. At the end of my basic training, they took away the 45-caliber pistol they trained me to use because I was a physician, and according to the Geneva Convention, physicians are not to carry weapons. They also told me that I needed to be trained in triage. Prior to being in the Air Force, I had never heard the word "triage." I asked them, "What's triage?" They told me that, in some disasters, there may be so many people injured that the number of people who need medical care may exceed the medical personnel, equipment, or supplies that would be needed to completely provide medical care for all the injured people. Therefore, in order to use the limited available resources for the maximum benefit, the triage officer has to rapidly examine all the injured people and decide in which of four triage categories each injured person should be placed. According to my instructor, those labeled *"Immediate"* require immediate care, or they will die. In contrast, those who should be labeled as *"Minor"* do have an injury but do not need medical-surgical care because they will get better by themselves. Those who receive the label *"Delay"* do need medical-surgical care, but this can be delayed without danger of death or serious disability. Finally, those labeled as *"Expectant"* are so seriously injured that even with treatment, they are most likely to die, or their treatment would take so much time and/or supplies that two or more people labeled *"Immediate"* would die or be terribly disabled. The instructor also explained that this labeling is a continuous process so that the triage officer continually evaluates those who are injured and, when needed, alters their label. For example, somebody who has been labeled *"Delay,"* with time, can become *"Immediate."*

When I was in the service, in spite of practicing internal medicine for four years, I decided that I would apply for resident training in neurology. I decided to become a neurologist because I was always very interested in disorders of brain function, and learning about triage helped me to understand the basis of selective action-intention, the "when" system. The human brain has limited resources for programming movements, and the human body also has a limited ability to perform simultaneous actions. Therefore, the "when" triage process is important for selecting the actions that are needed. The triage of patients is a system for deciding "when" to treat. Just as there are four levels used when triaging patients during a disaster, there are also four "when" decisions. The following are the four "when" decisions:

1 *Action Initiation (When to initiate an action)*: Since the human brain has limited ability to both program and perform simultaneous actions, the successful athlete has to first decide when to act to achieve a relevant goal. For example, when to swing the bat at a pitch. If the action is likely to achieve a goal, then the action should also be initiated at the correct time.

2 *Action Inhibition (When not to initiate an action)*: In many sports, members of a team attempt to provide signals that will lead to an athlete on the opposite team performing an action that would interfere with that athlete or that athlete's team achieving its goal. For example, a defender jumping when a basketball player fakes a shot would be an example of a failure in the "when" action system. If an action cannot achieve a goal, or is unlikely to achieve a goal, this action should not be initiated in order to preserve limited resources for the initiation of an action that will help to achieve the desired goal.

3 *Persistence (When to continue an action or series of actions)*: When there is a disaster, the medical staff has to continue their treatments until all those people who are injured have been treated. After an athlete initiates some actions, they are often required to continue this action until they have achieved their goal. Some sports, such as running, swimming, skating, skiing, and crew, require continuous action, and in many other sports, repeated actions or series of actions are needed. This ability to continue a needed action or actions is called motor persistence.

4 *Action Termination (When to stop)*: If, during a disaster, a patient who is undergoing surgery dies, the surgeon should not continue operating on this patient but rather move on to the next patient. Similarly, if a treatment is not working, it should be stopped. In sports such as football, if the team with the ball repeatedly uses the same play, this play would be less likely to be successful. Similarly, if a play repeatedly causes a loss of yardage, it would not be good for the team to continue using this same play.

Action Initiation

As mentioned in prior sections, one of the best means of learning how the brain stores knowledge and program-specific activities is studying people with brain dysfunction. Therefore, in the next sections, the various types of action-intentional programming deficits that are caused by brain damage and how these functions may be related to athletic skills will be reviewed.

Neurological impairments can also be caused by defective brain development, and even in people with normal development, there are individual differences in the functioning of the systems that mediate different activities. For example, in the middle of the nineteenth century, Karl Wernicke reported that patients with injury to the back (posterior) and top part of their left temporal lobe could no longer understand or correctly produce words (Figure 3.1). Wernicke demonstrated that this

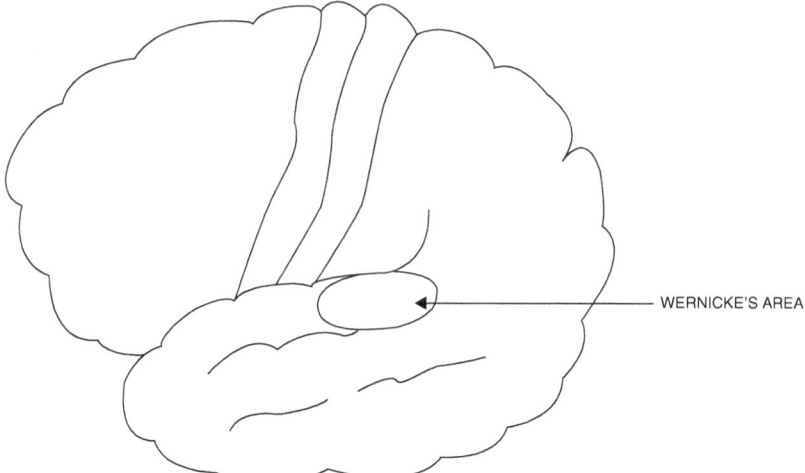

Figure 3.1 **Wernicke's area**. In almost all right-handed people and even in the majority of left-handed people, Wernicke's area is part of the auditory association cortex and is located in the back portion (caudal) and higher (superior) portion of the temporal lobe. Heard words are composed of a sequence of speech sounds (phonemes), and Wernicke's area stores the memories of previously heard words.

portion of the temporal lobe stores the memories of the sequence of speech sounds (phonemes) that compose the words that we have previously heard and learned. Thus, this part of the brain is like a sound-word dictionary. When this area of the brain is injured, patients cannot understand when other people speak to them; however, some of these people are able to comprehend written material. There are also patients who have injury to the left parietal lobe who have difficulty with reading but can understand speech. These patients can no longer read because the area of the brain that they injured contains the visual letter-word dictionary. There are other people who have similar disorders, but instead of having a disorder caused by injury such as stroke, they have developmental disorders that can cause similar problems. An example of this disorder is developmental dyslexia, where people have difficulty learning to develop this letter-word dictionary and are unable to learn to read. However, having a disability in one domain does not mean that other functions mediated by the brain are impaired. For example, there have been many great dyslexic athletes, and one of our heroes, here at the University of Florida, is Tim Tebow, who was a great quarterback and won the Heisman Trophy. He speaks wonderfully and has even been an announcer, but he is also dyslexic. These dissociations in skills reveal that the capacity of one system in a person's brain does not predict the capacity of other modules.

Athletes, as well as coaches and trainers, may want to be aware of and examine for limitations or weaknesses in the action-intentional systems that might reduce the ability of an athlete to be successful. These possible deficits or weaknesses can be categorized into the four broad categories which are mentioned earlier and include action initiation, response inhibition, persistence, and action termination.

Deficits in Planning and Action Initiation: Executive Dysfunction

Akinesia and Abulia

As mentioned earlier, the morpheme "a" means without, and "kinesia" means movement. Patients with akinesia do not initiate movements. There are several types of akinesia. For example, there can be global akinesia, where the person does do move any part of their body; limb akinesia, where they do not move a limb; directional akinesia, where they do not move the limb or eyes in a certain direction; and even hemispatial akinesia, where they fail to move in one half of body centered space. In addition, akinesia can be limited to just self-initiated movements. For example, patients with severe Parkinson's disease who are not taking their medicine may sit in a chair all day, but if a nurse comes in and says, "Let's take a walk together," they often will get up and walk with this nurse. Clinicians call this "akinesia paradoxica."

Since akinesia is not relevant for athletes, it will not be further discussed; however, there is a mild form of akinesia called abulia. The word abulia comes from two morphemes: "a," meaning without, and "bulia," which in Greek means "will." Hence, abulia is the loss of will or motivation to perform goal-oriented actions. Since most people do not have a complete loss of motivation to perform actions, perhaps a better term for abulia would be hypobulia, a reduction in motivation to perform activities. People with hypobulia have a decrease in the frequency of self-initiated, goal-directed activities.

While in some sports, actions are predominantly required as a specific response to a stimulus, like batting in baseball, in other sports, such as soccer, basketball, or hockey, there is a greater need for self-initiated movements. Some athletes could improve their performance if overall, they increased the frequency of movements. For example, during basketball games, after a team has the ball stolen while on offense, there may be one or two players who will not run to the other side of the court and go on defense.

Many lay people call this reduction of initiation, laziness. As explained earlier, to be a good or great athlete, a person must train, and often, in many sports, this training is self-initiated. Some people will not train no matter what the conditions; however, more often, a person with abulia will not initiate training by themselves but will train if ordered to do so by the coach or if this is requested by a fellow athlete (hypobulia paradoxica).

One of the first observations about the relationship between the frontal lobes and abulia came from the famous report about Phineas Gage, reported by John Harlow in 1868. Gage was a hardworking and energetic foreman of a crew that was laying down railroad tracks. An iron rod was used to place explosives in rocks. The explosive detonated while the iron rod was being used by Gage, and it became a missile that entered Gage's skull and his brain. This rod was so heavy, and its speed was so great that it even exited his skull. However, the rod seriously injured Gage's frontal lobes. It was amazing that he not only survived, but also, after this injury, he was able to move all parts of his body and speak. However, the people who knew him well reported that he was no longer Gage. Before this accident, he was highly energetic and persistent. After this accident, in spite of making plans for future actions, soon after making these plans, they were abandoned. It appears that this injury to his frontal lobes caused him to be abulic.

Karl Kleist (1934) studied soldiers who had frontal lobe injuries during the First World War and found that they often had a loss of drive and initiative, similar to the behavior Gage demonstrated. In addition, patients with neurological diseases who have injury to their frontal lobes or its connections with structures deep in the brain including the basal ganglia and thalamus also often display abulia. Neurologists and neuropsychologists often call the signs and symptoms associated with frontal lobe dysfunction an "executive disorder." They use this term because it is the executive that develops plans and puts these plans into effect.

Wally Nauta, in 1971, gave one of the clearest explanations of why patients with frontal injuries have a loss of goal-oriented behavior (abulia), especially those actions that are self-initiated. As reviewed in the introduction, the human cerebral cortex contains four major divisions or lobes (Figure 1.2), including the frontal, parietal, temporal, and occipital lobes. The parietal, temporal, and occipital lobes each receives sensory input. The parietal lobes receive somesthetic input (touch and temperature of items touching the skin as well as the position of joints). The temporal lobes receive auditory input, and the occipital lobes receive visual input. These primary receiving areas analyze the sensory input and then transfer this information to their modality-specific sensory association cortices, which are in the same lobes. These sensory association cortices synthesize incoming information. They also store the sensory memories of stimuli that were previously perceived. These modality-specific sensory association areas send this information to areas of the brain that integrate this information. These are called polymodal areas and are in other areas of the parietal and temporal lobes. This integration helps to provide meaning to the items we see, feel, and hear. For example, when the auditory association cortex in the left temporal lobe hears the letter sounds (phonemes) that form a word such as "shower," and this information is sent to polymodal cortex, that person may imagine a shower head with water coming out, hear the sounds of the water hitting the bathtub, and get the sensation of the warm water falling on their skin. This and other related information will allow them to know the meaning of the word shower and how a shower cleans the body.

There is another network in the brain called the limbic system, and this system is important in allowing us to feel emotions and make changes in our body that are produced by these emotions. For example, in the front middle part of each temporal lobe is a nucleus called the amygdala. When this amygdala is stimulated, the person may experience the emotion of anger and have the body changes associated with anger. The frontal lobes also connect with the hypothalamus that monitors our internal milieu and mediates biological drives such as hunger and thirst.

According to Nauta, the frontal lobes appear to fuse the information from these networks. This consolidation and fusion of drives, with emotions and knowledge of how to satisfy these drives, as well as how to reduce the situations that will give rise to negative emotions and enhance those that will give rise to positive emotions and reward, permit the frontal lobes to have the knowledge required by an executive. This frontal lobe executive network also has connections with cells deep in the hemisphere called the basal ganglia and disorders of these basal ganglia, such as Parkinson's disease, are often associated with deficits of initiation. Finally, the frontal lobes are connected with the premotor cortex, and the premotor cortex programs the motor cortex. Therefore, these connections enable the frontal cortex to initiate goal-oriented behaviors.

Planning

Athletic actions can be initiative or responsive. In general, offensive actions are more likely to be planned, and defensive actions are more likely to be responsive but not totally. Defense also has to be somewhat planned. Therefore, although the executive system is important in action initiation, even before an action is initiated, it is often planned, and the frontal executive system discussed earlier is critical for planning. Patients with frontal dysfunction caused by injury to the frontal lobes, as well as its multiple connections to other networks, not only have an impairment in self-initiation but often do not make plans or prepare. Even if they do make plans, and if their plans are unsuccessful, they will continue to make the same plans and take the actions, a disorder called perseveration that is described in a later section.

Hypokinesia

Hypokinesia is defined as the slowness in the initiation of goal-oriented movements. In many sports, the time taken to initiate a movement is critical for success. Hypokinesia is different than bradykinesia, which is slowness in the actual movement. Clinically, hypokinesia is most often caused by a decrease in the neurotransmitter dopamine, which is caused by diseases such as Parkinson's disease; however, hypokinesia can also be seen in patients with frontal lobe dysfunction.

As mentioned, when the nerve cells in the motor cortex fire (discharge), they send down messages through the brain and spinal cord to the nerve cells in the spinal cord, and the nerves in the spinal cord activate the muscles that initiate movement. In the resting state, these neurons have an electrical charge that is maintained by having different chemical concentrations inside and outside the neuron. When there is a sufficient change in this electrical charge, the neuron fires (action potential) and sends an electrical signal down its branches (long-branched axons and short-branched dendrites). The speed at which the motor neurons in the motor cortex can fire in response to a signal can be altered by changes in the resting voltage, and this change is performed by an area in front on the motor cortex called the premotor cortex. The premotor cortex becomes activated when it gets the correct signal from the sensory association cortex or from the prefrontal lobes.

The reaction time is the time it takes from the onset of a stimulus, such as light, for a person to press a button. Research has shown that when a warning stimulus comes before the signal to start an action, the person is able to start this action more rapidly than when there is no warning stimulus, and this saving is because the premotor cortex is preparing the motor cortex to fire. The effect of this preparation can be seen in many sports. If there is a race, and the starter states, "Take your mark...Get set...Go!" the runner or swimmer will be able to initiate an action more rapidly than if the starter just states, "Go."

In almost all sports, speed of initiation is important, and when there is no clear external signal to prepare for an action ("Get set!"), it would be valuable for the athletes to learn if there might be a hidden signal, such as the center raising the head before hiking the ball.

Reward

Since frontal function is dependent on input from many areas of the brain, the function of this executive is highly dependent upon connectivity. Many of the nerve cells (neurons) that are important for connectivity have the long cables called axons (Figure 1.1). Like wires that have insulation, these axons have a cover called myelin. Without this cover, these axons do not work as well. Many of the axons that interconnect with the frontal lobes do not get fully covered with myelin (myelinated) until a person reaches their early twenties. Therefore, many younger athletes may be less likely to self-initiate goal-directed activities and thus must often be supervised. One of the methods used by many coaches and teachers is reward.

Reward Network

Winning, scoring, and performing actions that have good results will often bring feelings such as satisfaction and joy to the athlete. Knowledge of these rewarding feelings not only motivates the athlete to initiate athletic activities but also are a driving force to train and practice so that future rewards can even be greater.

One of the first discoveries about the brain networks that provide the rewarding emotions of joy, pleasure, and satisfaction came from the work of James Olds and Peter Milner (1954). These investigators inserted electrodes into a region of the rat's brain near the nucleus accumbens. Humans have a similar area (Figure 3.2). These investigators put a lever in the cage of these rats so that when they pressed this lever, an electrical current was sent down this electrode to this region. These rats found that pressing this lever was so rewarding that they preferred to continue pressing this lever rather than eat or drink. Based on these observations, these investigators suggested that the area they stimulated is the pleasure or reward center of the brain.

There are two areas of the midbrain that contain neurons that make the neurotransmitter, dopamine. One area called the substantia nigra (Figure 3.3) sends dopamine to the basal ganglia of the brain. The loss

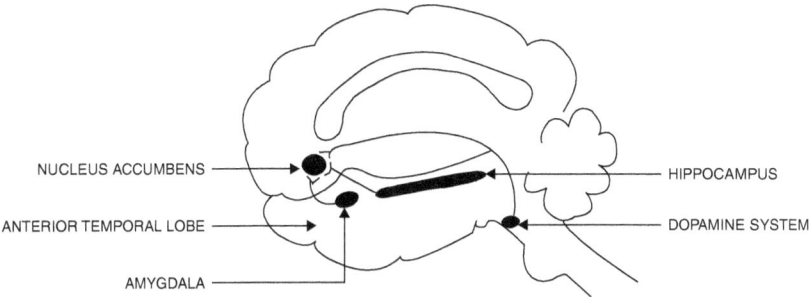

Figure 3.2 **Nucleus accumbens.** The nucleus accumbens is located in the basal forebrain. The basal forebrain is on the back (posterior) of the lowest (inferior) portion of the frontal lobes. The nucleus accumbens has connections with the amygdala (important for emotion) and the hippocampus (important in storing episodic memories). Stimulation of the nucleus accumbens provides a strong feeling of reward. This feeling of reward comes from activation of the dopamine system that has its neurons in the ventral tegmental area of the midbrain.

of these neurons causes Parkinson's disease with impaired initiation of movements, slowness of movements, shaking, and stiffness.

Another area called the ventral tegmental area sends dopaminergic neurons to this nucleus accumbens, and studies have revealed that when these neurons give off dopamine, it activates the nucleus accumbens and provides the animal the sense of reward. It is thought that addictive drugs, as well as food, water, and success, also cause the release of dopamine in the nucleus accumbens. Perhaps that is why for so many people, performing athletic activities, like drugs, is addictive, but athletic addiction is healthy.

Reinforcement

As a teenager, I heard an interesting fictional story about reward. A Jewish tailor emigrated from Germany to America, shortly after Hitler came to power in Germany. After coming to America, he looked for towns to open up his tailor shop. He found a town in Idaho that had a good size population but no tailor shops. He decided to move to this town and open his shop. He did not know that this town contained a very strong Nazi party with many members who weekly met in their meeting hall.

After this Jewish man opened his shop, and the members of this party found out about it, they made many anti-Semitic signs. It was summer, and the children were not in school, and so they asked the older children to go to this tailor's store and march around with these signs and shout anti-Semitic statements. About eight children decided to do this. After

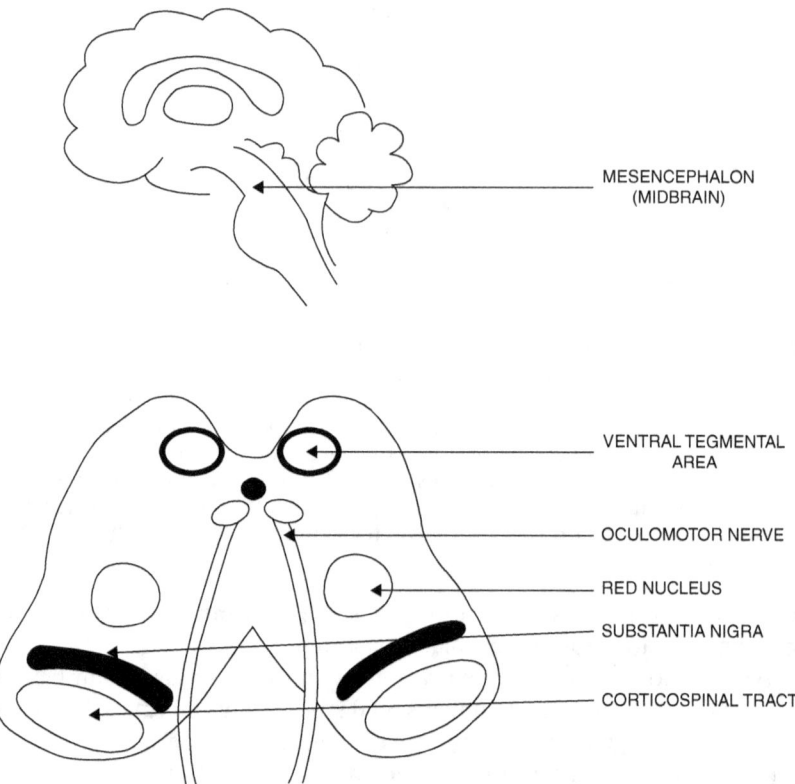

Figure 3.3 **Substantial nigra and ventral tegmental area.** The substantia nigra is in the midbrain. The figure on top is a cartoon of the brain cut in half from front to back and demonstrates the location of the transverse section through the midbrain. The lower figure is this transverse section of the midbrain showing the substantia nigra, the ventral tegmental area.

they marched around his store for about two hours, he came out of his store and gave each one of the children a five-dollar bill, and while handing this money to them, he said, with his Yiddish accent, "good job." When these children got back to the meeting hall, they all agreed that they did not understand why he did this, but they all decided that they would march around every day until he closed his store.

The next day, after they marched, he gave them four dollars, the following day three dollars, until his reward was only five cents. The following day, the children came to the Nazi meeting hall but decided not to march. When one of the Nazi leaders asked them why they were not marching they told him, "It is not worth marching for two hours for just a nickel."

B.F. Skinner, a famous experimental psychologist, first showed in experimental animals that as long as you reinforced animals with a reward such as food, they would continue performing a task, but shortly after the rewards ended, the animals stopped performing the task. However, if the animals received a random reinforcement, and this reward was discontinued, they would not stop performing this action, even without reward, for a long time. If you want to learn if this is also true for humans, I suggest you go to a gambling casino and watch people play the slot machines. Although these people are losing money, they repeatedly put their money into these machines and pull the handle. They continue putting their money into these machines because randomly they win a few coins.

Frontal lobe-mediated, goal-directed behavior is one of the most important factors for a person becoming an excellent athlete. Unfortunately, it is not clear how this can be trained. Most often, parents, teachers, and coaches try reward. Winning a sports event can be very rewarding and losing punishment. Perhaps engaging in sports is a means of enhancing a person's frontal executive functions; however, performing noncompetitive sports where there is no reward might work the best.

Response Inhibition

Response inhibition is characterized by withholding an action or actions in response to stimuli when these responses are not relevant for achieving a specific goal or even may interfere with achieving a goal. For example, during a football game, when the running back makes a fake action, and the defender purposefully withholds a response to this fake action (response inhibition) and instead continues to move in the actual direction of the runner and tackles the runner. However, some athletes are better at inhibiting responses than others.

Although defective response inhibition is almost always evoked by a stimulus, there are many means by which this incorrect response can be elicited, and there are several forms of defective response inhibition. For example, defective response inhibition may be seen in a specific part of the body such as the eyes (e.g., looking in the wrong direction), the legs, or the arms. The stimuli that most often induce a defective response are visual and tactile.

Studies of patients who show defective response inhibition often reveal injury to their frontal lobes. Denny-Brown and Chambers (1958) noted that all animals from the most primitive to the most highly developed have two major modes of activity: approach and avoidance. At the highest level, in the humans, the frontal lobes are responsible for disengagement and avoidance. In contrast, the posterior parietal and temporal lobes are responsible for engagement and approach. Patients with injury to their frontal lobes often reveal many abnormal approach behaviors. When touched on the palms, they will automatically grasp the object that touched their hand, called the grasp reflex. When they are seated at a table with objects in front of them, they will use and play with these objects. For example, if there is a glass and a pitcher of water in front of them, even if not thirsty, they will fill the glass with water and drink.

The frontal lobes do not fully mature until a person reaches about the age of 25. In addition, in many sports, the frontal lobes are often injured. Defective response inhibition is also often associated with attentional deficit-hyperactivity disorder (ADHD). Thus, there are many athletes who may reveal defective response inhibition and therefore be easily "faked out."

The athletes with ADHD should, of course, be referred for evaluation and treatment. There are also several exercises that may be used, such as Luria's 2/1 test. In this test or exercise, the participant is told that when they see the instructor hold up one finger, they are to hold up two fingers, and when the instructor holds up two fingers, they are not to hold up any fingers. Another task is the crossed response task, where the instructor holds up their left hand in the participant's right visual field and their right hand in the participant's left visual field. The participant is told to fixate their gaze on the instructor's nose, and when the instructor moves their right hand, the participant is to look at their left hand and vice versa. An even more complex task is to instruct the participant that when the instructor's hand moves upward, they are to look at the moving hand, and when it moves downward, they are to look at the opposite hand. If the participant delays their response, the task becomes easier, and thus, the participant should be instructed to respond as rapidly as possible.

I do not know if these tasks can improve response inhibition of athletes, but I have observed that when this task is repeated, many people will perform better. In addition, there are no studies that have determined if training on tasks such as these will generalize and lead to improved response inhibition during athletic events.

Motor (Action) Impersistence

If someone were to ask me what personality characteristic I thought was the most important for success independent of the domain, I would respond, "Persistence."

In order to be successful, a person has to not only initiate goal-oriented activities and avoid performing activities that do not lead to this goal but often to achieve a goal requires either a continuous action, repeated actions, or the continuation of a series of actions. When a person discontinues a goal-oriented activity before it is completed, this deficit is called motor impersistence.

A disorder of motor persistence was described by C. Miller Fisher (1956), a neurologist at Mass, General Hospital, who examined patients by requesting that they keep their eyes closed and tongue protruded.

He found that some patients who had a stroke could not maintain this posture for more than a few seconds. Subsequently, Andrew Kertesz and his coworkers (1985) suggested that the addition of mouth opening makes Fisher's test even more sensitive. When Kertesz and his coworker examined the location of the strokes that caused this impersistence, like the other intentional disorders that were described, they found that injury to frontal lobes, and especially the right frontal lobe was critical.

As mentioned, the frontal lobes fuse the information from networks that mediate drives, emotions, and knowledge of how to satisfy these drives as well as give rise to positive emotions and rewards. As also mentioned, the frontal lobes are connected with the premotor cortex and the premotor cortex programs the motor cortex, and these connections enable the frontal cortex to not only initiate goal-oriented behaviors and avoid behaviors that will not be rewarding or even harmful but also persist at behaviors until successful.

Success in all sports requires persistence, but in many team sports, like baseball and basketball, rewarding results, such as making a basket, are often accomplished shortly after an action is initiated. However, more prolonged persistence is important until the game is over. There are other sports that do not provide short-term rewards. In sports like track and swimming, especially at longer distances, persistence is critical. Although like many of the other behaviors reviewed, it is not entirely clear how to train persistence, running or swimming for longer distances may be a means of training persistence, as well as being excellent means of conditioning the body; however, like other training paradigms, it is best to start off doing what is easy and then progressing.

Motor Perseveration

There are several forms of motor perseveration including: (1) the continuation of an action or a series of actions after the task is completed; (2) the continuation of an action when it appears that this activity is not allowing or not helping a person to achieve their goal. The first type Sandson and Albert (1987) called "terminal motor perseveration," and the second, they called "flawed perseveration." Sandson and Albert (1987) described a third type called "recurrent perseveration," which is the repetition of a prior action after the task requirements have changed.

In the clinic, we test for motor perseveration by several means. In one test, the examiner asks the patient to copy his/her actions and performs a series of three hand actions described by Luria (1966). When seated, the examiner softly hits his own lap with his fist, then hits his lap with the ulnar edge of his open hand, and then finally hits his own lap with the palm of his open hand. Patients with recurrent perseveration may use their fist followed by the open hand and then again the fist, which would be recurrent perseveration; however, these errors may be related to an impaired action working memory deficit. Alternatively, the examiner may ask patients to draw a series of alternating triangles and squares or write the alternating letters of "n" and "m," and the patients with perseveration will repeat one shape or one letter.

Luria (1966) assessed patients for continuous perseveration by having them draw a triple loop. Patients with the continuous form of motor perseveration may continue to draw loops even after they completed drawing the triple loops. We have found that Dubois et al.'s (2005) clapping test is often a more sensitive test for detecting continuous perseveration. We ask the patient to copy what we are doing. Then, we clap our hands, about one clap per second, and after three claps, we stop for three seconds and, for second time, clap three times at one per second. If a patient claps more than three times they have perseveration.

It is unlikely that athletes will show the type of perseverations mentioned earlier, but athletes often do perform perseverative actions. Most often, but not always, their perseverative action is recurrent rather than continuous; however, sometimes, athletes do perform continuous perseveration. When an action is successful, athletes will often repeat this action either sequentially or intermittently; however, as the opponent learns about this action, it may become unsuccessful, and if the athletes continue to perform this unsuccessful action, it is perseverative.

To my knowledge, there are no exercises that can be performed to reduce perseveration except feedback, and that would be the job of the coach; however, sometimes, even coaches are perseverative.

4 Handedness

In many sports, hand-arm preference (handedness) is an important factor. There are many definitions of handedness, but most of the assessments for handedness examine preference rather than performance. For example, the questionnaires used to determine hand preference often ask many questions about the hand the person uses for a variety of activities. For example, they ask questions such as "With which hand do you write? What hand do you use to cut food with a knife? What hand do you use when you throw a ball?" Based on these questionnaires, people are classified as being right-handed, left-handed, or ambidextrous. In the general population, about 90 percent of people are classified as right-handed, about 8–9 percent are left-handed, and 1–2 percent are ambidextrous.

Handedness appears to be most important in those sports where the competition is interactive, such as baseball. In those sports that are directly not interactive, such as swimming, running, and downhill skiing, hand preference does not appear to be as important. In sports that are not interactive, the percent of right- and left-handed athletes is about the same as in the general population, but in many interactive sports, the percent of left handers is often double the number of right handers in the general population. For example, in the general population, about 9 percent of people are left-handed, but in professional baseball, about 20–25 percent of the players in the major leagues are left-handed.

The reason there may be more left-handed people in interactive sports is not fully known, but it is possible that in certain sports, left-handedness may provide the athlete with an advantage. For example, in baseball, it has been posited that if a person is right-handed and is up at bat, they stand on the left side of home plate. The left-handed batter stands on the right side of home plate. Therefore, the left-handed batter is closer to first base than the right-handed batter. Furthermore, when the left-handed batter swings his bat, she or he is more likely to be moving toward first base than the right-handed batter, who, when swinging, moves toward third base. Therefore, unlike the left-handed hitter, the right-handed hitter has to stop his momentum and change direction; this difference in position and direction of movement gives the left-handed batter a better opportunity to more rapidly get to first base. However, some studies that have examined this difference in timing and based on the results of these studies claim that this timing difference cannot fully account for the increase in the percentage of left-handed baseball players.

As noted by Poling et al. (2011), one of the rules about hitting in baseball is "bat right-handed against left-handed pitchers and bat left-handed against right-handed pitchers." Studies have found that baseball hitters do have higher success of getting on base when they are up against pitchers who have different handedness, and although there is a higher percentage of left-handed baseball players, there are still five times as many right-handed baseball players than left-handed, including pitchers.

Poling et al. (2011) also studied three great baseball players, Mickey Mantle, Pete Rose, and Eddie Murray, who had great batting averages and who were switch hitters (could bat as a left hander or as a right hander). These three great hitters, according to Poling et al., rarely batted right-handed against right-handed pitchers and rarely batted left-handed against left-handed pitchers. For example, during their entire careers in the major leagues, both Mickey Mantle and Eddie Murray batted right-handed against a right-handed pitcher only one time. Pete Rose batted left-handed against a left-handed pitcher three times. During all these five occasions, these three great batters never got a hit.

Another possible reason is familiarity. Even with the greater percentage of left handers in interactive sports, when compared to the general population, the percentage of right-handed athletes that participate in interactive sports is still much greater than those who are left-handed. In almost all interactive sports, recognizing the preparatory and early actions of the competitor will inform the athlete what she or he may expect. For example, knowing the windup of a pitcher in baseball may allow the batter to better prepare for the type of pitch that will be delivered, and this knowledge may allow the batter a greater probability of hitting the ball. In addition, when a left-handed pitcher throws the ball, it takes a different angle toward the batter than does the ball thrown by the right-handed pitcher. For example, when a right-handed pitcher throws a curve ball, it curves away from a right-handed batter but toward a left-handed batter, and it is claimed that it is easier to hit a ball coming toward than away from the batter.

Since there are more right-handed than left-handed pitchers, being a left-handed batter has an advantage, and since there are more right-handed than left-handed batters, a left-handed pitcher has an advantage over a right-handed pitcher. However, third basemen, shortstops, and second basemen are rarely left-handed, since they would have to rotate more than right handers when throwing the ball to first base. In the outfield, however, handedness makes little difference.

In tennis, watching the position of the opponent's body and arm during the back swing and the forward swing may allow the tennis player to have some knowledge of where the ball is likely to go as well as the speed of the ball and even if it will have spin and bounce irregularly. Since there are so many more right handers than left handers, when the athlete learns the meaning of these body movements and how to best respond to them, they are most likely learning from right-handed opponents, and thus when they compete with left handers, they may not recognize these movements as well as their opponents.

The third possible reason for an increase in the percentage of left handers in interactive sports may have to do with the degree of laterality. In sports such as baseball, many of the athletes have to use two hands: one to catch and one to throw. In combat sports such as boxing, both hands often have to be used. In addition, in many sports, the two hands have to be used together. For example, in baseball, when batting, both hands are used, and in basketball, shots are often performed with both hands. Many basketball players dribble with either the left or right hand, and on defense, the player often tries to block the ball with the hand that is closest to it. When right-handed people are given a handedness questionnaire, usually in response to almost all the questions, they will respond by answering "right." In contrast, left-handed people will answer "left" for the majority of questions, but in general, their percentage of left-hand responses is often lower than the right hander's right-hand responses. Thus, left handers appear to be somewhat more ambidextrous than right handers, and the ability of left handers to use their right hand is often better than right handers' ability to use their left hand. This difference may be another factor that allows more left handers than right handers to be great athletes.

The reason humans evolved to have hand preference is not entirely known; however, almost all mammals have a limb preference. There are at least three major reasons for the evolutionary development of hand-limb preference. First, many activities performed by people only require the use of one hand, with hand preference; one hand and arm can increase its skills with practice. Therefore, often, one hand/limb can perform activities better than the other. Second, in order to perform all forms of navigation, including walking, there must be an asymmetry within the navigator or an asymmetry of the environment. Many things in our environment are symmetrical. For example, when walking or driving a car, when we come to a fork in the road, there may be a sign telling us which direction to go. Or if you have previously driven on this road, when you come to the fork, there may be a certain building or a tree on one side of the road, and this structure may provide you with information about the direction in which you want to go. However, if there were no signs, and it was nighttime, and thus no asymmetrical environmental guides could be seen, how would you know which way to go? Since you have been on this road before, you may recall that you have to take the fork on the right side of the road. However, if anatomically and functionally, you were entirely symmetrical, you would not be able to know your right from left, and there would be a 50 percent chance that you would take the incorrect fork. Thus, for many people, their ability to know right from left on both themselves and in the environment depends upon hand preference. When I was a young boy, and someone asked me to turn to the right, I figured out which direction was right by asking myself, "Which hand would you use to throw a ball?" One of my children has a little mole on her left arm. When she started school and needed to know her left from her right, she would search for this mole, and when she found it, she knew that was her left arm. Thus, when we must determine right versus left and there are no environmental cues, we have to use our own body asymmetry.

The third reason for hand preference is related to hemispheric specialization. Each hemisphere is specialized to perform certain functions. In the latter part of the eighteenth century, Gall, the founder of phrenology, made two postulates about the brain's organization. Although he did not use these terms, he posited that the brain is organized in a modular fashion, such that different anatomic areas in our brains mediate different functions. His second postulate was that the larger and more developed this module, the better it would perform its functions. It has been known that the growth of the skull is dependent on the growth of the brain. Thus, according to phrenologists, they should be able to determine the size of the modules in the brain by measurements of the skull, and if they know the size of the module, they may be able to predict a person's ability to perform different mental functions.

Unfortunately, phrenology became a pseudoscience, with practitioners making many claims that were not substantiated by well-designed studies; however, in the mid-nineteenth century, the famous physician Paul Broca, who was interested in anthropology, went to a lecture being given by one of Gall's disciples. At this lecture, he learned that phrenologists believed that it was the frontal lobe that mediated speech. Broca knew about a patient in the hospital who had an impaired ability to communicate with speech. Although this patient could comprehend speech, he had such severe problems expressing himself that the only word he could say was "tan." This patient was in the hospital because he had gangrene of his lower extremity, and he eventually died. Broca examined this patient's brain after he died and found that he had a lesion centered in left frontal lobe. Subsequently, Broca collected a series of patients who had aphasia and died. He found that all these patients had injury to their left hemisphere. Based on Paul Broca's report, Gall's hypothesis that the brain is organized in a modular fashion, with different parts of the brain performing different functions, was supported. Several years after Broca's reports, another neurologist Karl Wernicke described patients who unlike Broca's patients had fluent but abnormal speech and were not able to comprehend other people's speech. Unlike Broca's patients, these patients had injury to the posterior portion of their left superior temporal lobe cortex (Figure 3.1), but again, their injury was to their left hemisphere.

Based on these studies of patients with aphasia, it was thought that handedness was determined by the hemispheric laterality of language. Many of the activities that we perform are based on verbal reasoning, and since the left hemisphere is critical for verbal reasoning and the left hemisphere also controls the right hand, hand preference was thought to be related to hemispheric dominance of language. For example, if a right-handed baseball pitcher recognizes the player who is up at bat and recalls that he struck out this batter the last time he pitched to him by throwing curve balls, and the pitcher decides to throw a curve ball, his verbal reasoning and memory would be mediated by the left hemisphere. Since the left hemisphere motor system also controls the right arm and hand, his brain could more directly get this message to the motor systems of the left hemisphere than to the motor systems of the right hemisphere that control the left hand. Similarly, since the left hemisphere of most right-handed people mediates language, writing with the right hand would permit more direct access to language representations than writing with the left hand.

The right and left hemispheres are connected by a large cable called the corpus callosum, and the corpus callosum allows the two hemispheres to communicate (Figure 1.6). This language laterality-handedness hypothesis is in part supported by patients who have a hemispheric disconnection after an injury to their corpus callosum. Several years ago, I had the opportunity to examine a woman who had her corpus callosum cut to prevent the epileptic seizures that were starting in one hemisphere from spreading to the other hemisphere. Like other patients with callosal disconnection, shortly after her surgery, she had several experiences that made her feel like her left upper limb was being controlled by alien forces. For example, she told me a story that once when she was getting dressed, after putting on her dress, she went to the closet to get shoes that matched her dress. She picked up these shoes up with her right hand, and her left hand pulled the shoes out of her right hand and reached for another pair of shoes that were a different color. Since she verbally reasoned that the color of the shoes that her left hand selected did not match the color of her dress, her left hemisphere did not want her to wear these shoes with this dress. So, she had her right hand take these shoes from her left hand and threw them on the floor. Shortly after doing this, the left hand slammed the closet door on her right arm. This is an example of why, under these types of circumstances, the left upper limb of a person with a callosal disconnection is often called an "alien limb."

In 1962, Norman Geschwind and Edith Kaplan described a right-handed patient who was undergoing surgery for removal of a brain tumor, and the surgeon obliterated an artery that carries blood to his corpus callosum; therefore this patient sustained damage to his corpus callosum. To learn if with his injured callosum he could transfer language messages from his left to his right hemisphere, they verbally asked him to carry out commands with his right and left upper limbs. With his right hand, he was able to correctly perform these movements, but he was unable to correctly make these movements with his left hand. Although he could not correctly carry out movements with his left hand to command, he could correctly imitate the examiner's movements with his left arm and hand. Therefore, it appears that his callosal injury prevented the verbal commands comprehended by his left hemisphere from reaching his right hemisphere's motor networks.

Although the laterality of language processing to one hemisphere may have an influence on handedness, it cannot explain why in about 70 percent of people who are left-handed, it is their left hemisphere that is dominant for mediating language. In addition, even in 5 percent of right handers, it is their right hemisphere that mediates language.

As described earlier, in order to perform learned skilled actions, such as those performed by a baseball pitcher or quarterback throwing a pass, a person has to know which joints to move, which not to move, how rapidly to move each joint, and how much force is needed. Joint movements are performed by muscle contractions, and it is the motor nerves coming from the spinal cord that provide the signals that make the muscles contract. These nerves in the spinal cord are activated by nerves coming from the motor cortex in the brain; however, the motor cortex must be provided with instructions from other parts of the brain that have stored the knowledge of how the different parts of the motor cortex should be activated.

In 1920, Hugo Liepmann reported that right-handed people who have left hemisphere injury often could no longer correctly perform learned skilled activities with either upper limb. He called this disorder "ideomotor apraxia." We subsequently analyzed the errors made by patients with ideomotor apraxia and found that they often made several types of errors. They often failed to correctly place their fingers, hand, and arm in the correct posture. For example, when asked to show the examiner how they would slice bread with a knife, they would often keep their hand open and move the open hand like it was the knife. They would often move the incorrect joint or joints. For example, when making believe that they were using a screwdriver to put a screw into the wall, they would rotate their hand at their wrist rather than rotate their forearm at their elbow. Many skilled movements require the coordinated movements of multiple joints. For example, when throwing a ball, a person needs to first start in a posture where their arm is lifted at the shoulder; their forearm is bent (flexed) toward their upper arm; their hand is held backward (extended) at the wrist; and their fingers are bent (partially flexed), enabling them to hold the ball. Their throwing action starts when they move their arm downward at the shoulder joint and then start straightening their arm at the elbow, followed by wrist forward flexion and then opening the hand to release the ball. Patients with ideomotor apraxia are impaired and programming these coordinated movements. In addition, patients with ideomotor apraxia move these joints at incorrect speeds (Poizner et al., 1990). Liepmann (1920) thought that it was the left inferior parietal lobe (Figure 2.2) that stored the memories of how to correctly program these movements. Subsequent studies of patients with apraxia (Rothi et al., 1991) as well as functional imaging (Moll et al., 1998) studies have provided support for this hypothesis. Since these movement representations and programs are often in the left hemisphere of right handers, Liepmann (1920) also hypothesized that it was the hemispheric lateralization of these spatial-temporal movement programs that accounted for handedness.

Since these memories of movements are stored in the left hemisphere when a person attempts to make a learned skilled movement, these movement memories would have more direct access to the premotor and motor areas of the left hemisphere that control the right upper limb than to the motor areas of hemisphere that control the left upper limb. In addition, since these movement memories are stored in the left hemisphere of people who are right-handed, these right-handed people may be better able to learn new skills with their right upper limb than with their left upper limb. It is for these reasons that Liepmann thought that it is this asymmetrical storage of these movement memories that is the major factor in determining hand preference.

Liepmann (1920) as well as other investigators have noticed that when some people who are right-handed have a stroke that injures the left inferior parietal lobe, including the supramarginal gyrus (Figure 2.2), where these movement memories are usually stored, they are not apraxic. In addition, we have reported that there are people who are right-handed who became apraxic with right hemisphere strokes (Raymer et al., 1999). These observations suggest that the lateralized storage of these movement representations cannot fully account for hand preference.

As mentioned earlier, when Norman Geschwind asked a surgeon why he wore magnifying glasses, the surgeon replied that now surgeons use magnifying glasses so that they can see what they are doing. Geschwind said to me how remarkable it was that surgeons are able to make such precise movements that they need magnifying glasses to see these movements. As also mentioned, we test for hand-finger deftness by using the coin rotation test where people are asked to rotate a coin (nickel) as rapidly as they can between their thumb, forefinger, and middle finger. Most right handers can do this more rapidly with their right hand than left hand. In addition, we mentioned earlier that right-handed people who have left but not right hemisphere injury lose the ability to make deft movement of their left hand; however, with injury to the right hemisphere of right-handed people, they may lose deftness of the left hand, but they usually do not lose deftness of their right hand (Hanna-Pladdy and colleagues, 2002; Heilman and colleagues, 2000). Furthermore, studies of a patient with injury to the corpus callosum provided evidence that the premotor or motor cortex in the left hemisphere supplies information to the right hemisphere's motor network that is important in the control of deft movements of the left hand (Acosta et al., 2014).

The motor cortex, which is at the back (posterior) portion of the frontal lobes (Figure 1.5), contains the nerve cells that send messages to the spinal cord that control movements of the limbs. Studies of the motor cortex reveal that the area of this cortex that controls the movements of the fingers (called the hand bump) in right handers is larger in the left than in the right hemisphere, and since more neurons may mean better control, this asymmetry may, at least in part, account for the asymmetry in hand deftness (dexterity), allowing right-handed people the ability to make more precise, rapid, and independent finger movements.

Based on the previous discussion, handedness or hand preference may rely on several factors including laterality of language, spatial-temporal movement programs, and asymmetry of motor innervation. However, there may also be other factors. In a prior chapter that I wrote for a book called, "Cubs on Your Mind," I mentioned that when I was a boy learning to play baseball, I was aware that it may be better for me if I could bat like a left hander. But I had great difficulty and found holding the bat like a left hander felt so uncomfortable. I did not understand how this could be explained by hand preference because I am using both hands. However, recently I think I may have learned the answer. If you are right-handed, I'd like you to try a little experiment. Now what I want you to do is close your eyes, and imagine that immediately in front of you is a straight vertical line that is about a foot long, and the middle of this line is level with your eyes. Now take one hand and put it on the top of this line and the other hand at the bottom. Now open your eyes. Is your right hand above your left? If so, you are like the majority of right handers. When you hold a bat and prepare to swing at a pitch, which hand is higher? One possible reason for a right hander to hold the bat with the right hand higher than the left could be that when holding a bat in this manner, the dominant hand is closer to the working part of the bat that hits the ball. However, it is not only in sports that right handers hold their right hand higher than the left. If you are right-handed and you are holding a shovel to pick up some dirt, which hand is higher? For most right handers, it is again the right hand. However, in this case it is the left hand that is closer to the working end.

It is not clear if this right higher than left is learned or is genetically preprogrammed. Most of the vertical actions we perform during our life with one hand and arm are downward movements. For example, when we eat, we move our fork, spoon, or knife downward. When we use tools, such as a hammer, a handsaw, or drill, our primary actions are also downward. Although this downward basis is most likely related to gravity, during the performance of most of these actions, when the left hand is used, it is often held lower than the right hand. Thus, it is unclear if holding the right hand higher than the left is learned or related to some inborn functional brain asymmetry.

When a right hander holds a bat, not only is their right hand higher than the left, but it is also closer to the body. If you look at the arms of right-handed batters when they are prepared to hit a pitch, their right arm at the elbow is much more flexed than their left arm. The flexion asymmetry can be observed in many sports. For example, right-handed boxers keep their right hand closer to their body than the left hand when they are not throwing a punch.

The overall thesis that we have promoted is that the hand a person selects to use is based on the activities mediated by the contralateral hemisphere. It was explained that whereas the left hemisphere is specialized for many of the functions, the right hemisphere is also dominant for many functions. Since in right handed people, as we mentioned, it is the left hemisphere than mediates deft movements, one of the things that I did not understand is why people who are right-handed, who play string instruments, such as the violin, cello, bass, and guitar, primarily use their left hand and fingers to make the deft movements required to play the melody. Then I recalled that many years ago, I read a paper where they studied patients who had portions of their right or left temporal lobe removed to help control their epilepsy. After their temporal lobectomy, their musical abilities were tested. Brenda Milner found that the patients with right temporal lobectomy had an impairment in melody perception, and those with left temporal lobectomy had a deficit of rhythm. Perhaps the reason people who play string instruments use their left hands to play the melody and their right hands to activate the strings is because the right hemisphere is dominant for mediating melody and the left hemisphere mediates rhythm.

There are also some sports in which the left arm and hand play a critical role in performance. Right handers who compete with rifles, as well as archery, typically have their right hand closer to their body than their left hand; however, this posture may not be entirely related to the preferred posture of the right hand and instead may be related to the different skills of the two arms and hands. When firing a rifle, the most important skill is aiming, and it is primarily the left upper limb that is performing the aiming. The only thing the right hand needs to do is make a flexion movement to pull the trigger or, when using a bow and arrow, extend the tips of the finger to let go of the string. In most right-handed people, it is their right hemisphere that is dominant for mediating spatial reasoning and spatial computations, and it is the right hemisphere that controls the left hand.

When I was a boy and was given a baseball glove, I saw that the glove was for my left hand. I asked my brother Fred, why is the glove for my left hand. He said, "Kenny, you catch with your left hand so that you can throw with your right hand. If you had the glove on your right hand you would not be able to use it to throw the ball." While clearly, my brother was correct, there may be an additional reason. Catching a ball is heavily dependent on visuospatial analysis, and it is the right hemisphere that is dominant for performing this analysis.

Finally, although many of the reasons for hand preference have been reviewed, the reason for the predominance of right hand preference is still not known. There are however several theories. Since Charles Darwin's theory of evolution, with the proposal of survival of the fittest, people have searched for reasons why right-handedness may have offered a survival benefit. One of the theories has suggested that since ancient times, humans have used spears to hunt animals and thus probably used spears to fight each other. To protect themselves from an enemy's spear, these ancient humans probably also used some type of shield. If they were right-handed, they would have probably held the spear (and later a sword) in their right hand and shield in their left hand. During a spear or sword fight, the most rapid means of killing your enemy is to stab them in their heart. In almost all left- and right-handed people, the heart is on the left side of the chest. Thus, the warriors who carried their spear or sword in their right hand and their shield in their left hand would be better able to protect their hearts than those who carried the shield in their right hand. Since these right-handed warriors were better able to protect their hearts, perhaps they were better fit to survive.

Despite the tremendous advances in genetics and neuroscience, we still do not entirely know what causes people to be left handers. Genetics cannot fully explain hand preference, and even two left-handed parents are still more likely to have more right-handed than left-handed children. Geneticists claim that genetics appear to have only a small (24 percent) influence on peoples' handedness. The scientific name for left-handedness is "sinistral," and sinister means ominous, malevolent, alarming, and frightening. The reason for calling left-handedness sinistral is not entirely known but may be related to future disability, disease, and even death. During fetal development, the brain may be injured. With brain injury of one hemisphere, there may be an alteration of hemispheric dominance and hand preference. If 90 percent of fetuses were destined to be right-handed, and only 10 percent were destined to be left-handed, then there would be a greater percentage of fetuses who sustained brain injury who would be converted from future right handers to left handers than those who changed from future left handers to right handers. Therefore, having a child who prefers to use their left hand may be a bad or sinister sign. In many cultures, parents and teachers routinely attempted to convert left handers to right handers. Perhaps they attempted this conversion because being left-handed or sinistral carried a negative and malevolent stigma. However, when it comes to athletics, being left-handed appears to be benevolent.

5 Attention

Definition

In general, words that are used to describe mental activities, such as attention, appear to be well understood by the majority of people but are often difficult to define. One of the most important mental activities that athletes must perform is properly allocating their attention. Our brain receives more sensory information than it can fully process. In addition to the information coming from the outside world, there is the information stored in our brain that we may activate and process. Because there is more information coming into our brain and being activated within our brain than we can fully process, we have to select the information that we wish to further process; this selection process is called attention. We select the information that is most relevant to our long-term goals, our short-term goals, and our immediate needs. Our immediate needs are based on our emotions and biological conditions, such as hunger, thirst, pain, need for oxygen, need to evacuate our body's wastes, etc. These immediate needs usually take the highest priority. For example, if you are driving your car and need to immediately go to the bathroom, and you find a gas station, you may not attend to or be aware of the name of the gasoline they sell or even the color of the bathroom door. Similarly, if you think reading this book may be important to your career, you will probably be totally unaware of feeling your buttocks touching the chair or your feet touching your shoes until you read this sentence. After a while, as you continue to read, you again will be unaware of how your feet or buttocks feel. This process of removing your attention-awareness from a meaningless stimulus is called habituation. A new or novel stimulus will, however, usually initiate attentional awareness so that if you felt something crawling up your foot, you would immediately attend to this sensation and remove the ant from your leg.

In competitive athletic events, the correct allocation of visual attention is often critical to success. Distraction is the allocation of attention to an irrelevant stimulus, and in competitive sports, teams often attempt to use distraction to interfere with their opponents' success. For example, in baseball, when there is a runner on first base, even if this runner does not plan to steal a base, he may leave the base in order to distract the pitcher from focusing on the batter and his upcoming pitch. In football, occasionally, you will see a linebacker charging toward the offensive line without intending to break through this line but rather attempting to distract members of the offensive line.

Neuroanatomy of Attentional Networks

Studies of patients with focal lesions of the brain, as well as imaging studies, suggest that the posterior inferior parietal lobe is important for the allocation of spatial attention (Figure 2.2). This area gets sensory perceptual information from the visual, auditory, and tactile sensory association areas, and thus, it can monitor information coming into the brain. This area of the brain also receives information from the frontal lobe executive system that is important in the planning and implementation of goal-oriented behaviors, and from the cingulate gyrus, which is important in providing information about internal affairs, such as emotion and drives (Figure 5.1). Corbetta and Shulman (2002) posited that there are two anatomically and functionally distinct attention systems in the human brain. The more dorsal frontoparietal system was proposed to mediate the top-down volitional spatial allocation of attention to specific locations or features, and the ventral frontoparietal system was assumed to be involved in detecting novel stimuli. Although there have been some functional imaging studies that provide support for this hypothesis, there have not been many studies of patients with focal lesions that reveal that lesions in these different dorsal versus ventral loci of the parietal lobe

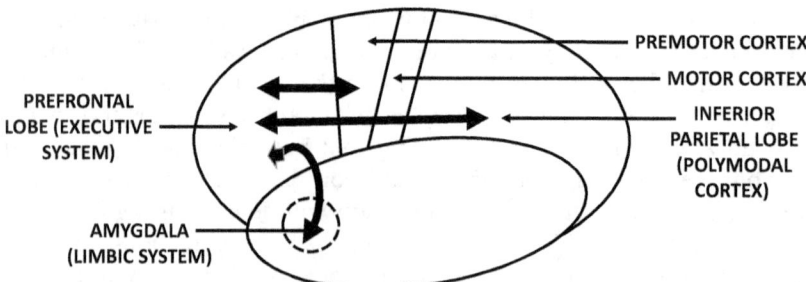

Figure 5.1 **Connectivity of the prefrontal area.** The prefrontal cortex, the cortical area in front of the premotor cortex, has reciprocal connections with many areas of the brain including the inferior parietal cortex, which is a polymodal area critical for cognitive functions, as well as action programing and the allocation of attention. The visual, auditory, somatosensory cortex also projects to the prefrontal lobe (not shown in the figure). This prefrontal cortex also connects with portions of the limbic system, such as the amygdala, important for mediating emotions, and the hypothalamus, important in the regulation of internal bodily functions.

produce different forms of inattention. However, Hillis et al. (1998) have revealed that more ventral lesions are likely to reveal allocentric neglect where patients are unaware of the left side of stimuli, independent of where they are placed in relation to the body, versus more dorsal lesions, which are egocentric and in which patients are unaware of stimuli on one side of their body and head.

Having two hemispheres in the brain is in part like having two brains that communicate. These two brains allow parallel processing. For example, when you listen to someone speak, the left hemisphere decodes the sounds that make the words, but at the same time, the right hemisphere decodes the prosody of the speech, which often conveys the emotion of the speaker. Sometimes, these two messages may conflict, and you may hear people say, "It is not what you said but how you said it." In regard to attention, there is some evidence that whereas the right hemisphere mediates global attention, the left hemisphere mediates focal attention. In addition, each hemisphere has a propensity to attend to the opposite side of space. If you want to see how this attention dichotomy influences your allocation of spatial attention, get a piece of blank white paper. Lay this paper on a table so that the longer part (11 inches) goes from right to left, and the shorter part (8 1/2 inches) goes up and down. Now, draw a line in the middle of the paper that is about nine inches long. Then, place the paper in front of you, and make a mark in the middle of this line (bisect the line). After you make your bisection mark, place the letter "R" on the top right of this sheet of paper so that you will know which side is on the right. Now, fold the paper so that the left and right ends of the lines are touching each other. Then, open it up. If you want to try this experiment, do it before you read the next paragraph.

When we attend to part of an object, the part that we attend to appears to be relatively larger than the part that is less attended to. When normal young people attempt to bisect the lines described earlier, in order to find the middle of the line, they must allocate global attention, and since it is the right hemisphere that allocates global attention and directs attention toward left space, most people, when attempting to bisect the lines, will often deviate a few millimeters to the left. However, please do not be concerned if you were more accurate or even deviated to the right. One of the reasons we may read and write from left to right is that before we initiate these activities, we first must allocate global attention to the page, and when allocating this global attention, we have a left-sided bias. However, when we read, we allocate focal attention mediated by the left hemisphere, and perhaps this is why we scan rightward. While driving, we also use global attention, and this may also produce a left-sided bias; perhaps this may be the reason we drive on the right side of the road. But this cannot explain why people who read and write in Semitic languages, such as Arabic and Hebrew, write and read in the opposite direction and why the British drive on the opposite side of the road.

In regard to sports, this global attentional asymmetry might suggest that the quarterback is more likely to see an open receiver on the left side of the field than on the right side. However, I do not know if anybody has tested this hypothesis. In addition, the running back may also be more likely to see an opening in the defensive line on the left side. In contrast, a safety or linebacker may be more likely to detect a runner that is going to their left than to their right. If this is correct overall, there should be a greater yardage gain with left-sided than right-sided runs and passes. But again, I do not know if this has been tested. In addition to football, there are many other competitive sports that require the allocation of global attention, and it may be of interest to learn how these left-global, right-focal attentional biases influence these other sports.

Vigilance

Another aspect of attention that may influence athletic performance is vigilance or the ability to sustain high levels of attention. When, as a boy, during baseball games when I was asked to play in the outfield and played right field, I was initially able to pay attention to the pitcher and batter; however, after a few innings, there were no balls hit to right field, and I would start to daydream. It was only after hearing the bat hit the ball that I would attend to the ball flying in the air. But first, I would have to find the ball, and since I did not attend to the pitch and the swing, I had less knowledge as to where the ball was going. Therefore, when the ball was hit to right field, and if it was not directly hit to where I was standing, I would have a late start, and I was not able to cover as much ground as I would have been able to cover if I had paid attention to the entire processes.

The ability to sustain attention over a period of time is called vigilance, and vigilance is another critical ability for athletes. People with attentional deficit disorder often have an impaired ability to sustain their attention. Patients with right hemispheric damage also have an impairment of vigilance, and studies of normal people indicate that the right hemisphere is dominant for mediating vigilance. Within the right hemisphere, it appears that again, the inferior parietal and frontal lobes are critical. In general, we are more vigilant when we are highly aroused than when we have a low arousal level. Arousal of the cerebral hemispheres is mediated by an area of the brain that is called the reticular-activating formation, and one of the most important parts of this reticular formation is found in our midbrain (Figure 5.2).

In order for the brain to correctly operate, neurons have to communicate with other neurons. As mentioned, the region where one neuron meets another is called the synapse. In order to communicate, one neuron gives off a chemical at the synapse called a neurotransmitter. Some neurotransmitters can excite the other neuron, and others can reduce the excitement of the other neuron, called inhibition. Some of the neurotransmitters that

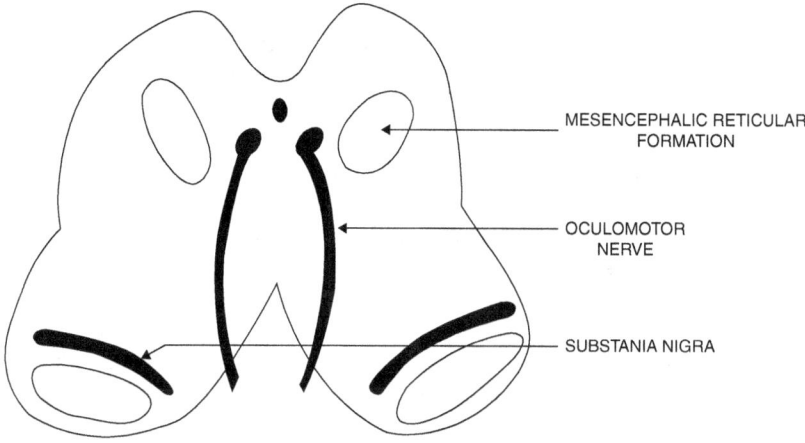

Figure 5.2 **Mesencephalic reticular formation.** This area of the midbrain plays an important role in activating (arousing) the neurons in the hemispheres, such that they can process incoming stimuli as well as program actions. When these areas in the midbrain are injured, the patient is comatose. The electroencephalogram (EEG) records electrical activity from the brain. When viewing an EEG, there appears to be waves similar to that seen in the ocean, and the more awake-aroused a person is during the recording, the more rapid the waves. The slower the waves, the less aroused and vigilant a person will be. We demonstrated that when a person damages his right hemisphere, and especially the inferior parietal lobe region, not only is there slowing of the EEG recorded from the injured hemisphere, but there is also slowing of the uninjured hemisphere. In addition, other studies, primarily in old world monkeys, have revealed that electrical stimulation of this area increases arousal as determined both behaviorally and by the EEG.

can excite are acetylcholine and norepinephrine. One of the transmitters that causes inhibiting is called gamma-aminobutyric acid or GABA. Certain medications taken by people that may help reduce anxiety or help them fall asleep are called benzodiazepines, such as Xanax, Librium, and Valium. These medications increase the brain's GABA. In contrast, other medicines that are used to increase arousal in people who have attention disorders, such as Ritalin, increase the brain's norepinephrine.

During an athletic event, there are many things that may reduce an athlete's arousal and vigilance, including a lack of sleep, certain drugs such as benzodiazepines, (e.g. Xanax, Librium, Valium), barbiturates, muscle relaxants (e.g. Baclofen and Soma), and recent head injury. Of course, when possible, all these should be avoided.

The knowledge of the importance of a competitive event can enhance arousal. Therefore, when performing almost all activities, if the athlete has low arousal, he or she is very likely to perform poorly. With an increase in arousal, performance often improves. For example, to study the effects of arousal on the motor system, Coombes et al. (2009) had healthy participants view unpleasant images and neutral pictures. They found that with these negative arousing images, the participants had increased activation of their motor system, as measured by a reduction of the time needed to initiate a response (reaction time) and the generation of a force. They also revealed that when viewing these images, the motor action potential recorded from the motor cortex also increased.

In regard to performance, an increase of arousal may be a double-edged sword. Yerkes and Dobson reported that the relationship between performance and arousal is a bell-shaped curve (Figure 5.3). Studies of arousal and performance have revealed that whereas performance improves with increasing arousal, there is a peak in the bell-shaped curve such that with further increases of arousal, performance progressively deteriorates.

The reason for this deterioration in performance with increasing high arousal is not entirely known; however, with high arousal, there is often anxiety. High arousal with anxiety may cause alterations in the allocation of attention. There are two major means by which people can allocate attention. When performing goal-related activities, a person can allocate their attention to the stimuli that will allow them to achieve their goals. As mentioned earlier, while reading this book, you may be unaware of feeling your feet touching your shoes. However, if a bug crawled into your shoe, you

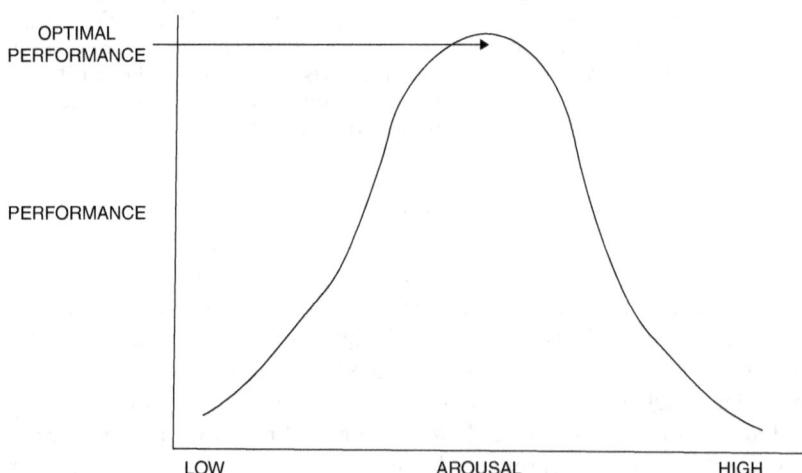

Figure 5.3 **Yerkes-Dobson curve.** The Yerkes-Dobson curve shows the relationship between arousal and performance. With low levels of arousal, i.e. when a person is tired, sleepy, and lethargic, their performance is poor, but as they become more aroused, their performance improves. With very high level of arousal, like that associated with high anxiety, performance deteriorates.

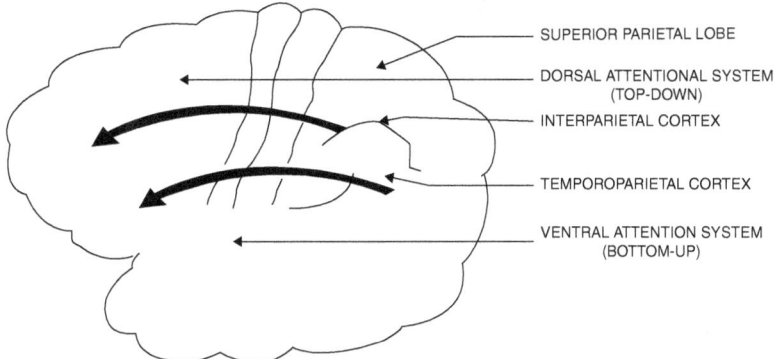

Figure 5.4 **Corbett's two attentional pathways.** Corbetta and Shulman (2002) posited that there are two different attentional networks. The more dorsal system, including the intraparietal cortex and superior frontal cortex, is responsible for goal-directed or top-down attention (e.g. searching the field for an open receiver). The other system is a bottom-up system that draws our attention to novel stimuli (e.g. an unseen defensive tackle appears from behind). This bottom-up system includes the temporoparietal cortex and inferior frontal cortex. In addition, the lower (ventral bottom-up) system works "as a 'circuit breaker" for the dorsal top-down system.

may then shift your attention to your foot. Goal-oriented attention is sometimes called "top-down," and the attention to novel stimuli "bottom-up." Corbetta and Shulman (2002) suggested that there may be two networks in our brains that carry out these two types of attentional functions. According to these investigators, one system, which includes parts of the intraparietal cortex and superior frontal cortex (Figure 5.4), is important for the "top-down" selection for stimuli. The other "bottom-up" system, which includes the temporoparietal cortex and inferior frontal cortex (Figure 5.4), is specialized for the detection of relevant stimuli that are novel-unexpected. When this "bottom system" becomes activated, it appears to inhibit the top-down system. During high states of arousal and with anxiety, there may be a shift so that the bottom-up system increases its activity. During completive sporting events, there are always novel visual stimuli occurring, and with high arousal anxiety, the athlete may be too easily distracted by novel stimuli, especially if these stimuli appear to be threatening (Bar-Haim, Lamy, Pergamin, Bakermans-Kranenburg, and van IJzendoom, 2007), and this distraction may decrease their athletic effectiveness.

6 Visual Perception

In almost all sports, vision and visual perception are critical. After light comes into the eyes and activates the cells in the retina, this information is sent back from the retina to the optic nerve (Figure 6.1). Near the pituitary gland, there is a structure called the optic chiasm, and it is at this location that the optic nerves divide. The portions of the nerve that come from the lateral sides of the retina in each eye do not cross, but the portions from the medial portions of the retina do cross sides in this optic chiasm (Figure 6.1). On each side of the brain, the optic track carries information from the lateral side of the retina on the same side and from the medial side of the retina on the opposite side. The retina is curved so that the lateral part obtains information from the opposite side of space that is across from each eye, and the medial retina obtains information from the same side of space. Therefore, each optic nerve carries information that comes from just one half of the area of space which a person is viewing. This visual information is carried to the visual thalamus that is deep in the brain (Figure 6.1). This area of the thalamus is called the lateral geniculate nucleus. This nucleus then sends this information to the primary receiving area in the occipital cortex that is on the same side. The left portion of the occipital cortex receives information from the portions of the retina that receive input from the right side of the viewing space and vice versa. The information from the center of the retina travels to the posterior (back) portion of the primary visual cortex, and the information from the more peripheral parts of the retina is carried to the more anterior (forward) portion of this cortex.

This primary visual cortex helps to analyze incoming visual information. The major function of the primary visual cortex is pattern recognition, and this pattern recognition is performed primarily by mapping the incoming information from the retina by way of the visual thalamus. From the primary visual cortex, these perceived patterns are then sent to the visual association cortex. The visual sensory association cortex, which is just in front of the primary cortex, can be divided into two major systems: one system that is in the lower part of the brain (the ventral system) and one that is higher up in the brain, the dorsal system (Figure 2.3).

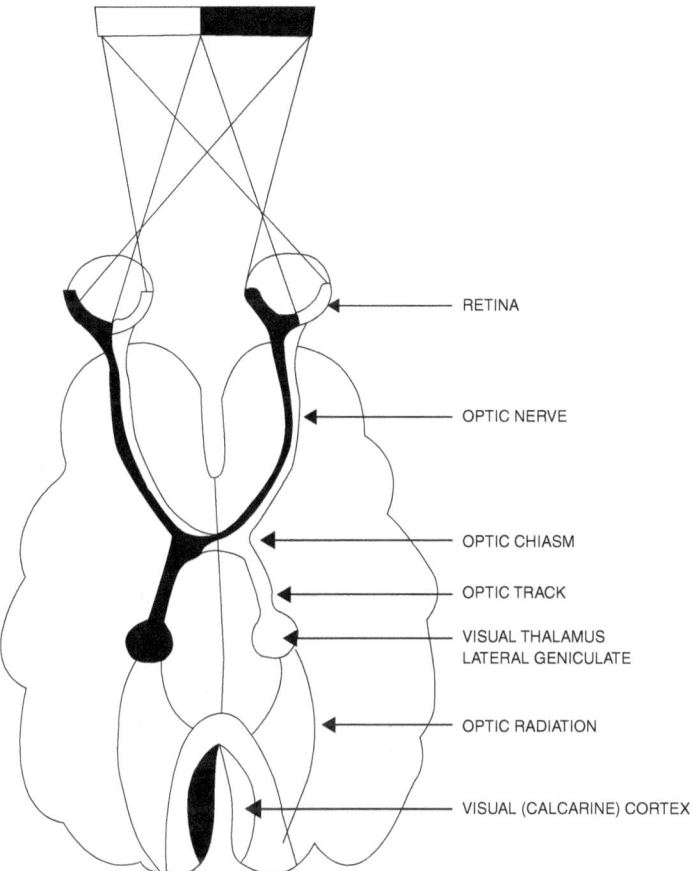

Figure 6.1 Visual pathways from the retina to primary visual cortex.

This lower system, which is at the bottom of the occipital and temporal lobes, stores the memories of those items that we have previously seen. On the right side of the brain, this visual association cortex stores images of the faces that we have seen. When the right side is damaged by diseases such as stroke, people have trouble recognizing peoples' faces even if they have seen these people thousands of times, but they may be able to recognize their voices. While these patients cannot recognize faces, they can recognize objects; however, if the left ventral visual association cortex is injured, the patients might not be able to recognize common objects. In addition, they may also have problems with reading written words. Ungerleider and Mishkin (1982) called this part of the visual processing network the "What" system. While athletes should be able to recognize their teammates' faces and objects such as a football, this ventral "What" visual system is not critical for athletes. Thus, if an athlete has poor facial recognition or even has problems with reading, it is unlikely to seriously affect his or her athletic performance. Similarly, an athlete who has great reading skills and is able to recognize many people's faces may not be a great athlete.

The upper portion of visual association cortex that includes the more dorsal (upper) portion of the occipital lobe association cortex as well as the parietal lobe appears to be important in performing the visual analyses that may be important in many sports. When this system is injured by a disease, such as a stroke, the patient has problems knowing where objects are in relation to her or his body. Thus, when such patients attempt to reach for an object with their hand, they will often miss this object and have to make many attempts to reach it. Bálint (1909) first described this disorder "optic ataxia." Bálint also reported that the patients with this disorder had a similar problem with the eye movements. Therefore, when they attempted to move eyes to focus on an object in their environment, their eyes would move to the incorrect location, and they would have to make multiple eye movements before they could focus on the object they wanted to view. A third problem that these patients have is a disorder call simultanagnosia. When patients with this disorder view a picture or their environment, they will focus on just one or two of the objects in the picture or in their environment. For example, several years ago I saw such a patient. Each year, a little north of Gainesville, they have a reenactment of a Civil War battle (Olustee). When I showed this woman a picture of this battle and asked her to tell me what this picture is about, she pointed to just one object and said, "Oh, I see a bird over here." When I reminded her to tell me what the entire picture was about, she then pointed to a horse and said, "I also see that horse." Thus, she was unable to put all the elements of the picture together. In order to be successful in many competitive sports, the athlete has to be able to move their eyes to the correct position, and they also have to be able to accurately reach for objects such as a ball. In many competitive sports, such as football, hockey, basketball, and soccer, the athlete also has to see the big picture, so they can plan their actions. Thus, this dorsal visual network appears to be very critical for athletes.

In many of the ball sports, it is also important to know where the ball is going. The distance a ball travels depends on several factors, including the speed of the ball and the angle of incline in relation to the athlete. In regard to speed, there are visual association areas called middle temporal (MT) and medial superior temporal (MST) (Figure 6.2) that are in the posterior portion of the temporal lobe. These areas appear to be important in the perception of motion. When this area was damaged in patients and they viewed actions, the motions of these actions did not flow and rather than appearing like a motion picture, these appeared like a series of still photographs. When I was playing right field, and finally a ball was hit in my direction, I ran to catch the ball. While I accurately

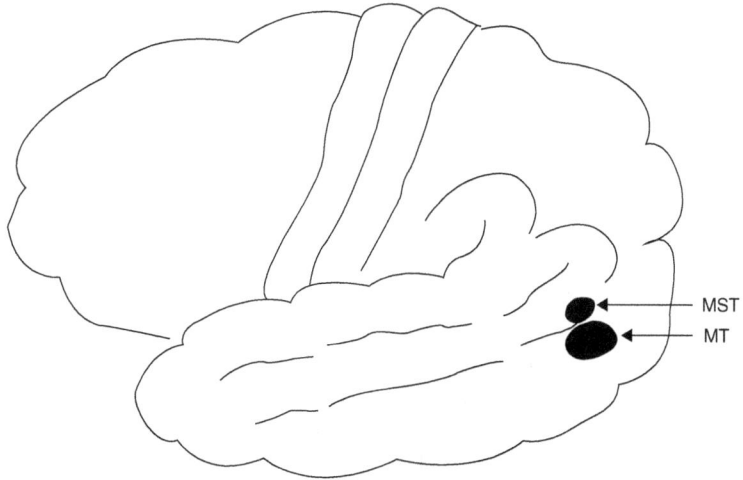

Figure 6.2 **Cortical areas MT and MST.** These are the two visual association areas that are important in the visual perception of movement.

ran to the correct right-left angle from home plate, I terribly misjudged the distance, and the ball went over my head. The next time I played, they asked me to play third base. Even now, I do not know how I would train someone to be better at having their brain compute, based on the upward angle and speed of a ball, where it would land.

7 Balance

As mentioned in the introduction, there are many sports in which balance is critical, such as skiing, gymnastics (e.g. balance beam), and surfing; however, in several other sports, while balance may not be as critical, it is often important. For example, in football, the running back often has to rapidly change the direction in which he is running, and to do so often requires excellent balance. In addition, the running back may get hit without being tackled, and a running back with good balance can continue to run.

The inner ear is not only important for hearing. In the inner ear, there is a structure called the cochlea that is important for hearing; however, there are other structures in the inner ear that are critical for balance, including the semicircular canals, the utricle, and the saccule. There are three semicircular canals that are approximately 90° from each other. Therefore, each semicircular canal is oriented to one of the three different spatial planes. When a movement takes place in any direction, the fluid in these canals moves and excites nerve cells that send messages to the brain about the direction of the movement. In addition, there are the utricle and saccule, which are important in detecting the head's position in relation to the force of gravity. Whereas the utricle is important in sensing horizontal movements, such as forward-backward movements and left-right movements, the saccule senses vertical, up-down movements. The utricle and saccule are able to detect these changes in position with their hair cells, which have little calcium stones (otoliths) on them and therefore are pulled by gravity. Thus, when the head changes position, these hairs bend, become activated, and send messages to the brain about this change in position.

Whereas the vestibular system plays a critical role in balance, this is not the only sensory organ that is important for balance. On one of my first visits to San Francisco, I wanted to see the Big Sur and rented a car to drive down to this beautiful park. On the way there, I saw a peninsula that extended out to the Pacific Ocean. I pulled the car over and off the road, got out of the car, and walked to the end of this peninsula, which was about two yards wide. It appeared that I was about 100 feet above the ocean, and I looked down, watching the wave coming in and going out. Then, I started to feel dizzy, like I had vertigo, and said to myself, "What a place to develop vertigo! You jerk! You can fall off this cliff and kill yourself." I figured that it would be safer to crawl back to the car on my stomach than risk walking and falling 100 feet. So, I crawled back to my car. Once I was back to my car, I stood up and found that I no longer had vertigo and asked myself, "What is going on?"

I recalled when I was a boy, and we were traveling in the Catskill Mountains, if someone sitting in the back seat got carsick, my mother would ask my father or my uncle Leon to stop driving for a while and then tell that carsick person to get in the front seat. Later, when I was sailing and someone who was below deck was getting seasick, I told them to go up to the deck and look forward at the horizon. We get seasick when the information from the vestibular system does not match what we are seeing with the visual system. When I was looking down at the water from the edge of the peninsula, it was the opposite of seasickness. When below deck in a boat, the vestibular system detects the boat rocking, but the visual system does not see this rocking. On the deck, a person can see the rocking. Similarly, in the back of a car, it is sometimes difficult to see all the turns the car is making, but from the front seat, the turns are more visible. When I was on this peninsula, I was watching the waves moving back and forth, but my vestibular system detected no movement, and it was this visual-vestibular mismatch that caused my vertigo. It also caused me to crawl, and when I got to my car, I was so covered with dirt that I headed back to the hotel for a shower and change of clothing. I never got to see the Big Sur. The point of this story is that an athlete's balance is not solely based on input from the vestibular system. The visual system is also critical. In addition, when our body moves, we also get feedback from the sensory nerves that are in our muscles, joints, and tendons. This sensory feedback system is called the proprioceptive system.

To have good balance, the information coming from these systems must be integrated as well as analyzed, and the output from this analysis has to influence the systems that control movement. The networks that mediate the integration of this information and program the movements required to maintain balance under a variety of conditions are not entirely known; however, there are several areas of the brain that, when injured, appear to impair balance and gait, including the vermis of the cerebellum and the mesencephalic gait center (pedunculopontine nucleus) (Figure 7.1).

There are many means of testing a person's balance and training balance. Many of these require machines, but others do not. One of the most commonly used exercises is standing on one leg first with eyes open and then, after balance improves, closing the eyelids. One of the best means of testing a person's ability to balance is the Star Excursion Balance Test (SEBT). Before the SEBT is performed, four lines (tape or chalk) should be placed on the floor. These lines should be about 8 feet long, and each

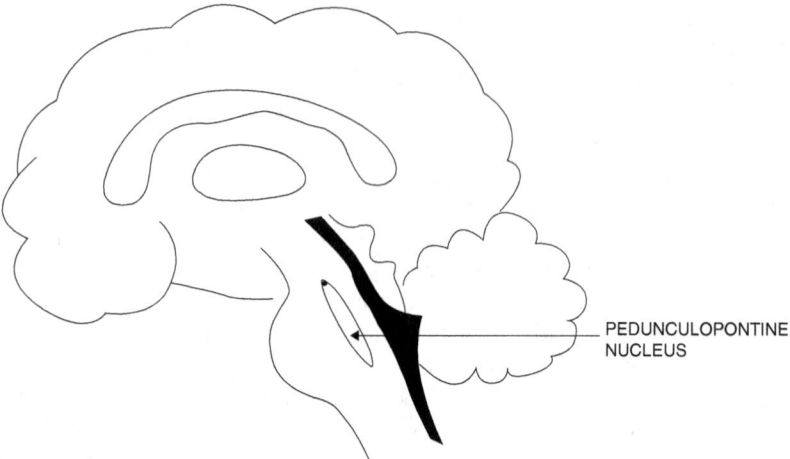

Figure 7.1 Pedunculopontine nucleus.

line should be 45° from the other line so that the lines form a star. The person performing the SEBT stands at the convergent point of these lines, lifts one leg, and places their leg in each of the eight positions, attempting to extend their leg so that it reaches as far as possible. When performing this test, the person must maintain balance on one leg. The person (standing on his/her left leg, for example) must reach in all eight different positions with their right leg. This test can also be used as an exercise.

Whereas it has also been shown that the performance of SEBT improves with training, to my knowledge, there have been no studies that have examined if athlete's performance improves after training using this test exercise. Recently, Kümmel et al. (2016) performed an extensive review of the literature with a meta-analysis of published articles between 1985 and 2015 to learn the influence of balance training of healthy people. Although they found that training on a specific task did improve performance on that task, in general, this improved performance did not generalize to other types of balance tasks. Based on the results, the authors concluded, "Consequently, therapists and coaches should identify exactly those tasks that need improvement, and use these tasks in the training program and as a part of the test battery that evaluates the efficacy of the training program."

8 Emotions and Mood

Anger

There are seven basic emotions, and these are happiness, sadness, anger, fear, disgust, surprise, and neutral. There have been many definitions of mood, but in general, many psychologists consider an emotion to be a more transient state often induced by a stimulus, while a mood is a more prolonged emotional state. For example, depression is a prolonged sad mood, and euphoria is a prolonged happy mood.

All emotions can be defined by three major elements: (1) valence (positive, negative, none) based on how the emotion makes a person feel (good, bad, or no feeling); (2) arousal, which is wakefulness-alertness; and (3) activity pattern (advance, retreat, none). For example, fear has a negative valence, high arousal, and avoid-retreat elements, and although anger also has a negative valence and high arousal, this emotion is associated with an approach-advance activity pattern.

An athlete's emotions can have a strong influence on their performance, and this influence can have positive and negative effects on their performance. One of the emotions that has the greatest influence on competitive athletes is anger. When I was a teenager, I occasionally boxed. Being right-handed, I was taught to stand with my feet apart and with my left foot in front of my right foot. In addition, I was instructed to keep my right hand closer to my chest than my left hand. I was told that the left hand was to be used to jab. The jabs were punches in which my left arm was rapidly extended to hit my opponent, and after hitting my opponent, rapidly flexing my arm, bringing it back to its starting position. My right hand was kept close to my chest until there was an opening in which it could reach my opponent's body or head. I was also taught that there were three types of punches that could be used by the right hand. When the other boxer was close to me, I was taught to use an upper cut, where the arm starts low and comes up; to use a straight shot when the opponent is an arm's length away; or to use a cross where the arm makes a curve. When making these punches, not only does the right-handed boxer use his more powerful right hand, but in the straight shot and cross, the boxer rotates his body so that there is even more force behind his punches.

78 Emotions and Mood

I usually was a pretty good boxer; however, when my opponent hit my nose with a jab, I would lose my temper and swing widely at my opponent and thus had trouble defending against my opponent's punches; that was not good because several of his punches would land on my head or body. Thus, anger made me more vulnerable because it made me not follow my strategic plans.

When I was training to be a neurologist, one of our Harvard neurosurgeons, Vernon Mark, presented a lovely young woman who was having fits of anger that were not provoked by anything she saw or heard. He found that she was having epileptic seizures that were coming from a nucleus in the brain called the amygdala, which is in the front part of the temporal lobe (Figure 8.1). When he removed this woman's amygdala, these episodes of anger stopped (Mark and Ervin, 1970). Subsequently, much research has been performed revealing that this nucleus is critical for producing emotions such as anger and fear. There is also evidence that the activity of the amygdala is controlled by portions of the frontal lobes. As we mentioned, the frontal lobes do not fully mature until people are in their twenties, and therefore younger athletes may have more difficulty controlling their anger.

There are several types of medication that can help control anger, but many of these medications have side effects that could interfere with athletic performance and should only be used as a last resort. First and foremost, it is important to avoid head injuries. Head injuries may impair the frontal lobe's ability to control the amygdala and shorten people's anger fuse. Second, it is important for teams to periodically inquire if the athletes are having anger control problems; however, many people will not admit to a loss of control. Next, if there is a loss of anger control, there

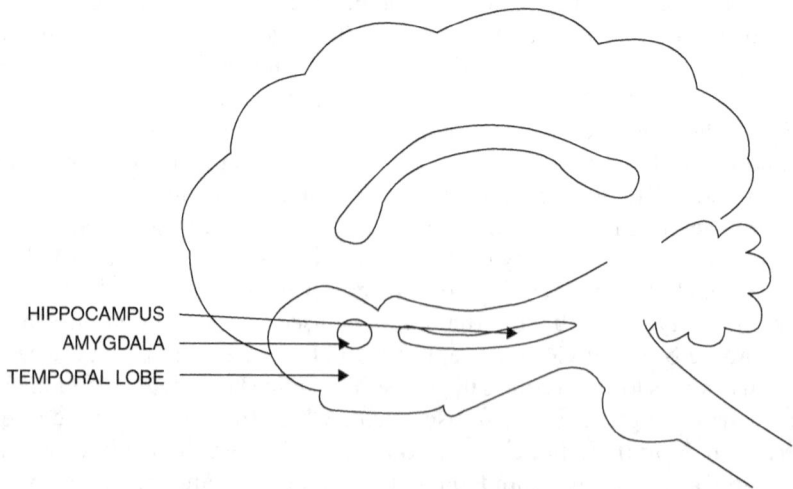

Figure 8.1 Amygdala.

Emotions and Mood 79

are a variety of self-regulation programs that are available, including educational programs where people are taught self-control strategies.

Depression

Depression is one of the major causes of human suffering. There are several major signs and symptoms that can be associated with depression, and these include: (1) depressed mood including feelings of sadness, grief, irritability, worthlessness, guilt, and emptiness; (2) a loss or lack of pleasure and loss of interest when performing activities; (3) a change in appetite with a significant change in weight; (4) alterations of sleep patterns, including an impaired ability to fall asleep or excessive sleepiness (insomnia or hypersomnia) or early morning awakening; (5) change in activity with agitation or fatigue with a loss of energy; (6) frequent thoughts about death or suicide.

There are several successful means of treating depression. Psychotherapy is often successful and there are many medications that can help. These include medications, such as tricyclic antidepressants (e.g. nortriptyline), serotonin reuptake inhibitors (e.g. sertraline), combined serotonin and norepinephrine uptake inhibitors (e.g. duloxetine), and monoamine oxidase inhibitors. In patients with severe depression who do not recover with medications and psychotherapy, electric convulsive therapy is often helpful. There is good evidence, however, that exercise helps reduce and prevent depression (Gullette and Blumenthal, 1996; North et al., 1990). Engaging in certain sports that require sustained physical effort can probably do the same. For example, playing sports like billiards or bowling does not require sustained physical effort and thus are unlikely to help reduce or prevent depression, but sports where there is much running, like soccer, lacrosse, and basketball, may not only provide aerobic exercise but also stop the athlete from having ruminating thoughts and brings joy and social engagement. In addition, when exercising on a machine, such as a treadmill, a person is more likely to discontinue exercising than when engaging in a sport.

In regard to the intensity of exercise that may relieve depression, there are still many unanswered questions; high-intensity aerobic exercise (65%–75% VO2 max reserve) does not appear to be much more of a help than less intense exercise, but to help depression, a person should exercise at least three times per week for more than a total of 150 minutes. It may take 10 to 15 weeks before a person experiences an improvement in their depression.

The reason exercise may help treat depressions is not known; however, it is possible that people who exercise develop a sense of mastery and improve their self-regard. Exercise may alter the neurotransmitters in the brain, increasing both serotonin and brain norepinephrine, which are decreased in people with depression. Norepinephrine is broken down by an enzyme called catechol-O-methyltransferase (COMT), and this enzyme is higher in patients with depression. Exercise appears to reduce the activity of this enzyme. Exercise also appears to increase endorphins and endocannabinoids. The term endorphins comes from two words: endogenous (within) and morphine. Endorphins, like morphine, inhibit the perception of pain and may even produce euphoria. Endocannabinoids are also produced by the brain and influence the perception of pain and improve mood.

9   Traumatic Brain Injury

Injuries are common in sports, including broken bones, joint injuries, bruises and lacerations of the skin, broken teeth, broken noses, and tears of muscles and tendons. However, some of the most serious injuries are those to the brain. The American Academy of Neurology defined concussion as an alteration of brain function manifested by a change in mental status caused by a mechanical force that can be associated with a loss of consciousness. There are several means by which a mechanical force can cause brain dysfunction. In sports, the most common causes are whiplash with rapid acceleration and deceleration, and direct skull percussion.

It has been estimated that in the United States, there are almost four million reported head injuries per year; however, many athletes do not report their injuries. The most frequent causes of injury are car accidents, but for teenagers and those in their twenties, athletic injuries are the next most common. These injuries are most common in those sports in which there is bodily contact, such as boxing, wrestling, football, hockey, soccer, basketball, and lacrosse. The sports with the least injuries are individual competitive sports, such as swimming, running, bowling, weight lifting, and billiards. In a recent study, James Noble reported at the American Academy of Neurology that women athletes are 50% more likely to have a sports-related concussion than male athletes. The reason women have a higher rate of concussion is not known, but it does not appear to be directly related to the types of sports played. For example, the women who played soccer and basketball were more likely to have a concussion than their male counterparts. Men or women athletes with history of a prior concussion were approximately three times more likely to have another concussion as those who had never had a concussion.

In the midbrain (Figure 5.2), there is a system called the reticular activating system, which is important in increasing and maintaining the activation of the neurons in our cerebral cortex. This system also allows us to be conscious. With head injuries, there may be a temporary loss of consciousness caused by a dysfunction of this system.

Traumatic injury to the head can cause several types of deficits. It can cause fractures of the skull, and these can be associated with a loss of spinal fluid and bleeding. Surrounding the brain is a skin-like membrane called the dura. With head injuries, there can be a hemorrhage below the skull and above the dura called an epidural hemorrhage or below the dura called a subdural hemorrhage. Hemorrhage can also occur in the substance of the brain. If a hemorrhage is large enough, it can push the brain to one side or down, which is called cerebral herniation. If this occurs, it can cause death or serious disability. These hemorrhages can be easily diagnosed with brain imaging, such as computer tomography (CT) or magnetic resonance imaging (MRI), and it is important that any athlete with a serious head injury has imaging, especially if they have problems with their thinking and memory as well as being less aroused.

Another problem that can occur with traumatic brain injury (TBI) is swelling of the brain (cerebral edema). This is more likely to occur in younger athletes, such as teenagers. However, this edema can be so severe that it can also cause brain herniation and death; therefore, it must be immediately medically/surgically treated.

Memory

The brain can also be bruised, which is called a contusion. These injuries often occur in the parts of the brain that are adjacent to bones that protrude. Two of the most common areas to be contused are the frontal lobes and the lower part of the temporal lobes.

One of the most common disorders associated with head injury is a loss of episodic memory. A simple test of episodic memory is the three-word delayed recall test. When giving this test, the athlete is given three unrelated words to remember (e.g. chair, elephant, and telephone) and asked to repeat them immediately. After he or she repeats all three of these words, the athlete's orientation should be tested. With a head injury, the athlete may not know the date, the day of the week, or the score before he or she got injured. After the questions about time orientation, the athlete should be asked to recall the three words. If the athlete is not fully oriented to time or cannot recall the three words, there is a high probability that the athlete has a disorder of episodic memory.

It is not entirely clear why head trauma causes an impairment in episodic memory. In 1957, Scoville and Brenda Milner reported the famous patient, HM, who had epilepsy that could not be controlled by medication. Electroencephalographic (EEG) records revealed that his seizures were coming from both temporal lobes, and so the neurosurgeon removed both of these. After his temporal lobes were removed, HM could not recall new episodic memories. Subsequently, Milner and Klein (2016) reported that it was the lower middle (ventromedial) part of the temporal lobe that was critical for storing these memories. The critical structure is the hippocampus (Figure 2.7). With head injuries, the ventromedial temporal lobes are often injured, and the loss of memory associated with TBI may be related to injury to the hippocampus.

Another area that is often injured is the frontal lobes, and whereas the hippocampus is important for storing memories in the cerebral cortex, the frontal lobes are important for the retrieval of memories. Many neurons in the cortex communicate with other neurons in the cortex that are some distance away, as well as other neurons deep in the brain, under the cortex. These long connections are mediated by the neurons' axons, which travel together and form cables called fasciculi. The frontal lobes are the area of the cortex whose function is most dependent upon these long connections. The frontal lobe has connections with the posterior portions of the brain with sensory association areas and polymodal areas (longitudinal fasciculi); connections with portions of the emotional limbic system, such as the amygdala (uncinate fasciculus); and connections with the thalamus and the basal ganglia. These long axons are easily damaged by shearing forces when concussions produce acceleration-deceleration forces as well as rotational forces. In addition, TBI also often causes many changes in the neurotransmitters that are important in communication between neurons as well as the cellular metabolism of the neurons and blood flow to these networks.

Executive Functions

In addition to being important for memory retrieval, the frontal lobes are important for executive functions. Executives perform the following critical activities: 1) planning ahead, 2) initiating activities that will enable these executives to accomplish their planned goals, 3) preventing and discontinuing activities that do not help them to accomplish their goals or that may cause harm, 4) continuing productive activities until the goal is accomplished, 5) stopping activities when the goal is completed or the activity does not appear to accomplish the goal. Therefore, people with frontal executive dysfunction are impaired at future planning and are impaired at initiating goal-oriented behaviors. As mentioned earlier, a decreased ability to initiate goal-oriented activities is called *abulia*. When there are activities that they are able to initiate, they often discontinue these activities before they have completed their task. This disorder is called *impersistence*. When there are stimuli that they see or hear that are not related to their goal, they are easily distracted (*distractibility*) and may engage in activities that do not help accomplish their goal (*defective response inhibition*). Even after they have accomplished their goal, or perform an activity which is not helpful in accomplishing their goal, people with executive dysfunction may continue these same actions. This disorder is called *perseveration*.

There have been several studies that have examined football players who have had multiple TBIs, and these studies have revealed that these retired players often reveal impairments of memory and executive function (Ford et al., 2013; Hart et al., 2013; Randolph et al., 2013). MRI studies of some of these athletes have revealed that these changes are often associated with evidence of injury to subcortical white matter and frontal executive functions are dependent on these white matter connections.

Emotions and Mood

The frontal lobe helps to control emotions and mood. With TBI, one of the areas often injured is the ventral (bottom) portion of the frontal lobes called the orbitofrontal cortex (Figure 9.1). This area connects to the temporal lobe and then the amygdala. The orbitofrontal cortex helps to control the amygdala that mediates anger. Therefore, people who play certain sports, such as football, in school or professionally, often have recurrent TBIs, and many of these injuries are likely to injure the frontal lobe and its connectivity to the amygdala. In general, the rule in the control of neuronal networks is that when an area of the brain gets disconnected, this area becomes hyperactive, and hyperactivity of the amygdala causes a low threshold for anger outbursts. Studies of soldiers with a history of TBI have revealed that a history of TBI increases the risk of problems with the experience, expression, and control of anger

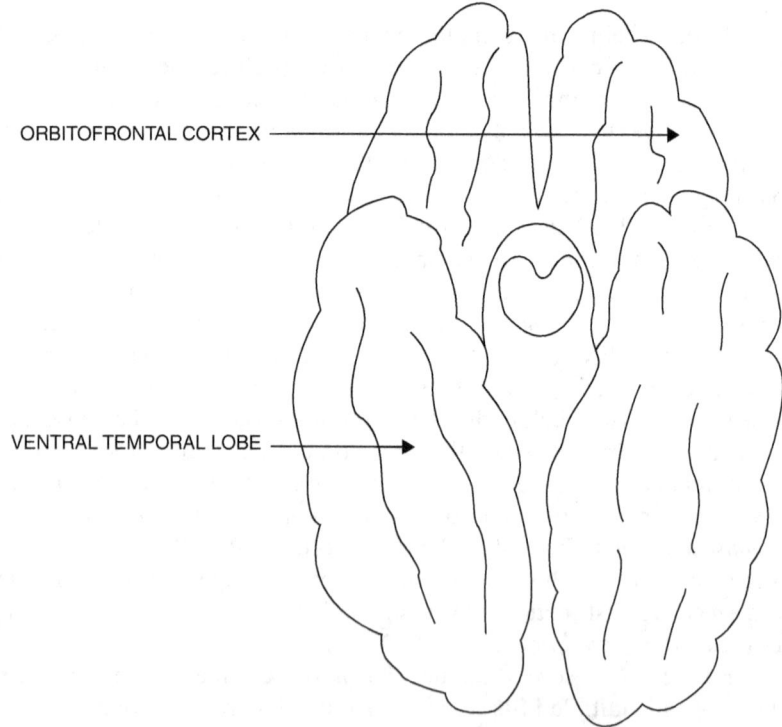

Figure 9.1 Orbitofrontal cortex.

(Bailie et al., 2015). There have been many stories of football players who became inappropriately violent and injured other people. For example, *Sports Illustrated*, on September 11, 2014, wrote that in the last two years, 33 National Football League players were arrested for violent acts.

Studies of patients who have had strokes that injure the frontal lobes, and especially the left hemisphere, reveal that these strokes induce depression. The reason for this is still not known. Although it has been proposed that it is the disability associated with the stroke that is causing this depression, strokes that just involve the temporal, parietal, and occipital lobes are not as likely to cause depression. Repeated TBIs also appear to alter mood and induce depression. For example, Guskiewicz et al. (2007) reported that retired football players with three or more concussions had a threefold increase in the probability that they would suffer from depression. Studies of these players have demonstrated that they have damage to the white matter that connects cortical structures with subcortical structures, such as the thalamus and basal ganglia, as well as those that link cortical structure with portions of the limbic system (Hart, 2013).

Chronic Traumatic Encephalopathy

The sport where athletes are most likely to experience multiple TBIs is boxing. The boxer who produces a serious concussion in his opponent, with a loss of consciousness for more than ten seconds, is often the winner: a knockout. In 1928, Harrison Martland published a paper in which he described boxers who developed a progressive deterioration of brain function that he called "Punch Drunk." Perhaps he called this disorder "Punch Drunk" because in many respects, these boxers looked like people who drank too much alcohol. When walking they would stagger, called "ataxia" and have slurred speech called "dysarthria." Their movements were also slow, and some also showed shaking, a "tremor." Subsequently, the name of this progressive disorder was changed to "dementia pugilistica," and then the British neurologist Critchley, in 1949, called this disorder "chronic traumatic encephalopathy."

Some of the major disturbances associated with this disorder are terrible emotional disorders, including depression, with suicidal inclinations, and inappropriate aggressiveness. In addition, people with this disorder also develop a progressive dementia.

The neurons in the brain have little tubes that supply information and other critical elements to their distant branches. There are proteins (tau) that are important for stabilizing these microtubules, but in diseases such as Alzheimer's, these proteins are altered, and this causes impaired function and eventually the death of neurons. McKee and her coworkers (2009) have revealed in postmortem studies of football players who suffered from chronic traumatic encephalopathy that the neurons in the cerebral cortex had evidence of pathological changes in their tau proteins. The cerebral cortex contains six layers, and McKee et al. found that tauopathy in patients with chronic traumatic encephalopathy affected Layers 2 and 3, and appeared to be arranged around small blood vessels. In contrast, patients with Alzheimer's disease have their tauopathy in Layers 3 and 5. These pathological changes in the cortex and degeneration of axons cause several changes in these athlete's brains, including atrophy of the cerebral cortex and subcortical white matter. This atrophy is very prominent in the frontal lobes, which are important in executive functions and emotional control as well as in the hippocampus, which is the area of the brain critical for encoding episodic memories. With time, this atrophy develops in other parts of the brain, and on postmortem examination, these athletes' brains were of low weight and had enlarged ventricles.

Until the brain is examined, chronic traumatic encephalopathy cannot be definitively diagnosed, and unfortunately, there is no specific treatment that can reverse or even stop the progress of this disease. However, prevention is better than a cure, and there are so many sports that do not cause repeated head injuries, it is not clear to me why high schools have football teams. Football makes much money for colleges but not for the athletes who are injuring their brains. About four decades ago, the largest association of neurologists, the American Academy of Neurology, tried to have boxing outlawed. After all, dogfighting was successfully outlawed, but not boxing. It is money rather than health that controls many sports.

10 Influence of Exercise on Age Related Cognitive Decline and Dementia

With aging, there is often a decline of several forms of cognitive functions. If this decline in cognitive functions interferes with a person's ability to care for themselves and to perform their job, then this person has a dementia. There are many forms of cognitive functions, including episodic memory (e.g. recalling what you had for dinner yesterday evening), working memory (e.g. recalling the telephone number the operator gave you), speech and language (e.g. word-finding difficulties), calculating (e.g. balancing your checkbook), performing skilled movements (slicing a turkey breast with a knife), being able to navigate (driving to a specific location), planning ahead and initiating goal-oriented actions.

With aging, one of the most common deficits is a decline of episodic (recent) memory. Studies have revealed that the critical area for storing these memories is a structure in the medial temporal lobes called the hippocampus (Figure 2.7), and with aging, this is one of the main areas that reveal atrophy. Erickson and his coworkers (2011) noted that the hippocampus and medial temporal lobe volumes are larger in adults who are physically fit. They also noted that physical activity increases the amount of blood that goes to the hippocampus. They, therefore, wanted to learn if aerobic training can increase the hippocampal volume in older adults and improve their episodic memory. In a randomized controlled study, some participants performed the aerobic exercises, and others performed non-aerobic tasks. They found that the participants who performed the aerobic exercise tasks experienced an increase in the size of the anterior hippocampus, with improvements in their memory. In the control group, the hippocampal volume and memory declined as it does with aging. These important findings indicate that aerobic exercise is effective at preventing and even reversing hippocampal volume loss in older adults as well as improving memory. In addition, other studies have demonstrated that moderate-intensity exercise appears to prevent the development of mild cognitive impairment (Geda et al., 2010). There is also evidence that aerobic exercises performed throughout adult lives reduce the probability of developing dementia, including Alzheimer's disease.

The reason exercise improves memory and even the growth of the hippocampus is not entirely known. However, the brain contains a brain-derived neurotrophic factor, and this factor appears to be like a fertilizer for neurons. There is evidence that aerobic exercise increases the level of this factor.

If your physician approves, exercise and play safe sports. You are likely to live longer, maintain a good memory, and enjoy life.

References

Acosta LM, Bennett JA, Heilman KM. Callosal disconnection and limb-kinetic apraxia. *Neurocase.* 2014;20(6):599–605.

Bailie JM, Cole WR, Ivins B, Boyd C, Lewis S, Neff J, Schwab K. The experience, expression, and control of anger following traumatic brain injury in a military sample. *J Head Trauma Rehabil.* 2015 Jan–Feb;30(1):12–20.

Bálint R. Seelenlähmung des "Schauens", optische Ataxie, räumliche Störung der Aufmerksamkeit (pdf). *Eur Neurol.* 1909;25(1):51–66, 67.

Bar-Haim Y, Lamy D, Pergamin L, Bakermans-Kranenburg MJ, van IJzendoorn MH. Threat-related attentional bias in anxious and nonanxious individuals: A meta-analytic study. *Psychol Bull.* 2007 Jan;133(1):1–24.

Bohlhalter S, Hattori N, Wheaton L, Fridman E, Shamim EA, Garraux G, Hallett M. Gesture subtype–Dependent left lateralization of praxis planning: An event-related fMRI study. *Cereb Cortex.* 2009 Jun;19(6):1256–62.

Bottjer SW, Johnson F. Circuits, hormones, and learning: Vocal behavior in songbirds. *J Neurobiol.* 1997 Nov;33(5):602–18.

Broca P. Perte de la Parole, ramollissement chronique et destruction partielle du lobe antérieur gauche du cerveau. *Bull Soc Anthropol.* 1861;2:235–8.

Coombes SA, Tandonnet C, Fujiyama H, Janelle CM, Cauraugh JH, Summers JJ. Emotion and motor preparation: A transcranial magnetic stimulation study of corticospinal motor tract excitability. *Cogn Affect Behav Neurosci.* 2009 Dec;9(4):380–8. doi:10.3758/CABN.9.4.380.

Corbetta M, Shulman GL. Control of goal-directed and stimulus-driven attention in the brain. *Nat Rev Neurosci.* 2002;3:201–15.

Coslett HB, Roeltgen DP, Rothi LG, Heilman KM. Transcortical sensory aphasia: Evidence for subtypes. *Brain Lang.* 1987;32:362–78.

Cotzias GC. L-Dopa for Parkinsonism. *N Engl J Med.* 1968 Mar 14;278(11):630.

Denny-Brown D, Chambers RA. The parietal lobe and behavior. *Proc Assoc Res Nerv Ment Dis.* 1958;36:35–117.

Dubois B, Slachevsky A, Pillon B, Beato R, Villalponda JM, Litvan I. "Applause sign" helps to discriminate PSP from FTD and PD. *Neurology.* 2005 Jun 28;64(12):2132–3.

Erickson KI, Voss MW, Prakash RS, Basak C, Szabo A, Chaddock L, Kim JS, Heo S, Alves H, White SM, Wojcicki TR, Mailey E, Vieira VJ, Martin SA, Pence BD, Woods JA, McAuley E, Kramer AF. Exercise training increases size of hippocampus and improves memory. *Proc Natl Acad Sci U S A.* 2011 Feb 15;108(7):3017–22. doi:10.1073/pnas.1015950108. Epub 2011 Jan 31.

References

Fisher CM. Left hemiplegia and motor impersistence. *J Nerv Ment Dis.* 1956;123:210.

Fogassi L, Gallese V, Buccino G, Craighero L, Fadiga L, Rizzolatti G. Cortical mechanism for the visual guidance of hand grasping movements in the monkey: A reversible inactivation study. *Brain.* 2001 Mar;124(Pt 3):571–86.

Ford JH, Giovanello KS, Guskiewicz KM. Episodic memory in former professional football players with a history of concussion: An event-related functional neuroimaging study. *J Neurotrauma.* 2013 Oct 15;30(20):1683–701.

Geda YE, Roberts RO, Knopman DS, Christianson TJ, Pankratz VS, Ivnik RJ, Boeve BF, Tangalos EG, Petersen RC, Rocca WA. Physical exercise and mild cognitive impairment: A population-based study. *Arch Neurol.* 2010 Jan;67(1):80–86.

Geschwind N. Disconnection syndromes in animals and man. *Brain.* 1965;88:237–94, 585–644.

Geschwind N, Kaplan E. A human cerebral disconnection syndrome. *Neurology.* 1962;12:65–685.

Goodglass H, Kaplan E. Disturbance of gesture and pantomime in aphasia. *Brain.* 1963;86:703–20.

Gould D, Udry E, Tuffey S, Loehr J. Burnout in competitive junior tennis players: A quantitative psychological assessment. *Sport Psychol.* 1996;10:322–40.

Gullette ECD, Blumenthal JA. Exercise therapy for the prevention and treatment of depression. *J Pract Psychiatr Behav Health.* 1996;5:263–71.

Guskiewicz KM, Mihalik JP, Shankar V, Marshall SW, Crowell DH, Oliaro SM, Ciocca MF, Hooker DN. Measurement of head impacts in collegiate football players: Relationship between head impact biomechanics and acute clinical outcome after concussion. *Neurosurgery.* 2007 Dec;61(6):1244–52.

Hanna-Pladdy B, Mendoza JE, Apostolos GT, Heilman KM. Lateralised motor control: Hemispheric damage and the loss of deftness. *J Neurol Neurosurg Psychiatry.* 2002 Nov;73(5):574–7.

Hart J Jr, Kraut MA, Womack KB, Strain J, Didehbani N, Bartz E, Conover H, Mansinghani S, Lu H, Cullum CM. Neuroimaging of cognitive dysfunction and depression in aging retired National Football League players: A cross-sectional study. *JAMA Neurol.* 2013 Mar 1;70(3):326–35.

Hebb D.O. (1949). *The Organization of Behavior.* New York: Wiley & Sons.

Heilman KM, Rothi LJG. (2012) Apraxia. In: *Clinical Neuropsychology*, 5th Edition. Eds. KM Heilman and E Valenstein. New York: Oxford University Press, pp. 214–37.

Heilman KM, Meador KJ, Loring DW. Hemispheric asymmetries of limb-kinetic apraxia: A loss of deftness. *Neurology.* 2000;55:523–6.

Heilman KM, Rothi LJ, Valenstein E. Two forms of ideomotor apraxia. *Neurology.* 1982;32:342–6.

Heilman KM, Schwartz HD, Geschwind N. Defective motor learning in ideomotor apraxia. *Neurology.* 1975;25:1018–20.

Heilman KM, Tucker DM, Valenstein E. A case of mixed transcortical aphasia with intact naming. *Brain.* 1976;99:415–26.

Hillis A, Rapp B, Benzing L, Carmazza A. Dissociable coordinate frames of unilateral spatial neglect: Viewer-centered neglect. *Brain Cogn.* 1998;37:491–526.

Kertesz A, Nicholson I, Cancelliere A, Kassa K, Black SE. Motor impersistence: A right-hemisphere syndrome. *Neurology.* 1985 May;35(5):662–6.

Kleist K. (1934). *Gehirnpathologie.* Leipzig: Barth.
Kühn S, Brass M. Testing the connection of the mirror system and speech: How articulation affects imitation in a simple response task. *Neuropsychologia.* 2008;46(5):1513–21.
Kümmel J, Kramer A, Giboin LS, Gruber M. Specificity of balance training in healthy individuals: A systematic review and meta-analysis. *Sports Med.* 2016 Sep;46(9):1261–71.
Lawrence DG, Kuypers HGJM. The functional organization of the motor system in the monkey. *Brain.* 1968;91:1–36.
Lichtheim L. On aphasia. *Brain.* 1885;7:733–84.
Liepmann H. Apraxia. *Ergbn Ges Med.* 1920;1:516–43.
Liepmann H, Mass O. Fall von linksseitiger Agraphie und Apraxie bei rechsseitiger Lahmung. *Z Psychol Neurol.* 1907;10:214–27.
Lissauer H. Ein Fall von Seelenblinheit nebst einem Beitrag zur Theorie derselben. *Arch Psychiatry.* 1890;21: 222–70.
Luria AR. (1996). *Higher Cortical Functions in Man.* New York: Basic Books.
Marcuse H. Apraktiscke Symotome bein linem Fall von seniler Demenz. *Zentralbl Mervheik Psychiatr.* 1904;27:737–51.
Marler P. Birdsong and speech development: Could there be parallels? *Am Sci.* 1970 Nov–Dec;58(6):669–73.
Mark VH, Ervin FR. (1970). *Violence and the Brain.* New York: Harper and Row.
Mateer CA. Executive function disorders: Rehabilitation challenges and strategies. *Semin Clin Neuropsychiatry.* 1999;4(1):50–59.
McCarthy R, Warrington EK. A two route model of speech production: Evidence from aphasia. *Brain.* 1984;107:463–85.
McKee AC, Cantu RC, Nowinski CJ, Hedley-Whyte ET, Gavett BE, Budson AE, Santini VE, Lee HS, Kubilus CA, Stern RA. Chronic traumatic encephalopathy in athletes: Progressive tauopathy after repetitive head injury. *J Neuropathol Exp Neurol.* 2009 Jul;68(7):709–35.
Mehler MF. Visuo imitative apraxia. *Neurology.* 1987;37:129.
Milner B, Klein D. Loss of recent memory after bilateral hippocampal lesions: Memory and memories-looking back and looking forward. *J Neurol Neurosurg Psychiatry.* 2016 Mar;87(3):230.
Mishkin M, Ungerleider LG. Contribution of striate inputs to the visuospatial functions of parieto-preoccipital cortex in monkeys. *Behav Brain Res.* 1982;6(1):57–77.
Moll J, De-Oliveira-Souza R, De-Souza-Lima F, Andreiuolo PA. Activation of left intraparietal sulcus using fMRI conceptual praxis paradigm. *Arq-Neuropsiquiatr.* 1998;56(4):808–11.
Nauta WJ. The problem of the frontal lobe: A reinterpretation. *J Psychiatr Res.* 1971 Aug;8(3):167–87.
Nirkko AC, Ozdoba C, Redmond SM, Bürki M, Schroth G, Hess CW, Wiesendanger M. Different ipsilateral representations for distal and proximal movements in the sensorimotor cortex: Activation and deactivation patterns. *Neuroimage.* 2001 May;13(5):825–35.
North TC, McCullagh P, Tran ZV. Effect of exercise on depression. *Exerc Sport Sci Rev.* 1990;18:379–414.
Ochipa C, Rapcsak SZ, Maher LM, Rothi LJ, Bowers D, Heilman KM. Selective deficit of praxis imagery in ideomotor apraxia. *Neurology.* 1997;49(2):474–80.

Oldenziel K, Gagne F, Gulbin J. (2004). Factors affecting the rate of athlete development from novice to senior elite: How applicable is the 10-year rule? Paper presented at the 2004 Pre-Olympic Congress—Sport Science through the Ages. Thessaloniki, Greece.

Olds J, Milner P. Positive reinforcement produced by electrical stimulation of septal area and other regions of rat brain. *J Comp Physiol Psychol.* 1954 Dec;47(6):419–27.

Orgogozo JM, Larsen B. Activation of the supplementary motor area during voluntary movement in man suggests it works as a supramotor area. *Science* 1979;206:847–50.

Pick A. (1905). *Sudien uber Motorische Apraxia und ihre Mahestenhende Erscheinungen.* Leipzig: Deuticke.

Poizner H, Mack L, Verfaellie M, Rothi LJG, Heilman KM. Three dimensional computer graphic analysis of apraxia. *Brain.* 1990;113:85–101.

Poling A, Weeden MA, Redner R, Foster TM. Switch hitting in baseball. *Exp Anal Behav.* 2011 Sep;96(2):283–9.

Randolph C, Karantzoulis S, Guskiewicz K. Prevalence and characterization of mild cognitive impairment in retired national football league players. *J Int Neuropsychol Soc.* 2013 Sep;19(8):873–80.

Raymer AM, Merians AS, Adair JC, Schwartz RL, Williamson DJ, Rothi LJ, Poizner H, Heilman KM. Crossed apraxia: Implications for handedness. *Cortex.* 1999 Apr;35(2):183–99.

Rothi LJG, Heilman KM. Acquisition and retention of gestures by apraxic patients. *Brain Cogn.* 1984 Oct;3(4):426–37.

Rothi LJG, Heilman KM. (1985). Ideomotor apraxia: gestural learning and memory. In: *Neuropsychological Studies in Apraxia and Related Disorders,* Ed. EA Roy. New York: Oxford University Press, pp. 65–74.

Rothi LJG, Mack L, Heilman KM. Pantomime agnosia. *J Neurol Neurosurg Psychiatry.* 1986;49:451–4.

Rothi LJG, Ochipa C, Heilman KM. A cognitive neuropsychological model of limb praxis. *Cogn Neuropsychol.* 1991;8:443–58.

Sandson J, Albert ML. Perseveration in behavioral neurology. *Neurology.* 1987 Nov;37(11):1736–41.

Schwartz RL, Barrett AM, Crucian GP, Heilman KM. Dissociation of gesture and object recognition. *Neurology.* 1998;50(4):1186–88.

Scoville WB, Milner B. Loss of recent memory after bilateral hippocampal lesions. *J Neurol Neurosurg Psychiatry.* 1957;20:11–21.

Thorpe WH. (1956). *Learning and Instinct in Animals.* Methuen, London.

Ungerleider LG, Mishkin M. (1982). Two cortical visual systems. In: *Analysis of Visual Behavior.* Eds. DJ Ingle, MA Goodale, RJW Mansfield. Cambridge, MA: MIT Press, pp. 549–86.

Verstichel P, Meyrignac C. Left unilateral melokinetic apraxia and left dynamic apraxia following partial callosal infarction. *Rev Neurol (Paris).* 2000;156(3):274–7.

Watson RT, Heilman KM. Callosal apraxia. *Brain.* 1983;106:391–403.

Watson RT, Fleet WS, Rothi LJG, Heilman KM. Apraxia and the supplementary motor area. *Arch Neurol.* 1986;43:787–92.

West MJ, King AP. Female visual displays affect the development of male song in the cowbird. *Nature.* 1988 Jul 21;334(6179):244–6.

Weirsma D. Risks and benefits of youth sport specialization: Perspectives and recommendations. *Pediatr Exerc Sci.* 2000;12:13–22.

Wildgruber D, Kischka U, Ackermann H, Klose U, Grodd W. Dynamic pattern of brain activation during sequencing of word strings evaluated by fMRI. *Brain Res Cogn Brain Res.* 1999 Jan;7(3):285–94.

Wyke M. The effects of brain lesions on the learning performance of a bimanual coordination task. *Cortex.* 1971;7:59–71.

Yamashita H. Perceptual-motor learning in amnesic patients with medial temporal lobe lesions. *Percept Mot Skills.* 1993 Dec;77(3 Pt 2):1311–4.

Index

abulia 42–5, 83
accuracy sports 11
action inhibition 40
action initiation 40, 41–6; akinesia and abulia 42–5; hypokinesia 45–6; planning 45
action-intention: action inhibition 40; action initiation 40, 41–2, 42–6; action termination 41; akinesia and abulia 42–5; hypokinesia 45–6; motor (action) impersistence 50–1; motor perseveration 51–2; overview 39–41; persistence 40; planning 45; response inhibition 49–50; reward 46–9
action-intentional programming deficits 41
action recognition 34–7
action sequencing 33
action termination 41
age: and birds' songs 27; and exercise 87–8; and motor learning 27; and sports 26, 28
aging, and Parkinson's disease 28–9
agnosia 34; pantomime 36
akinesia 42–5; abulia 43; directional 42; global 42; hemispatial 42; limb 42
akinesia paradoxica 42
Albert, M. L. 51
American Academy of Neurology 81
amygdala 44, 78
anger 77–9
apraxia 15; ideational 37; ideomotor 15–18; and motor acquisition defect 26
ataxia 11; optic 20
athletic actions: defensive 45; initiative 45; offensive 45; responsive 45
athletic events see sports
athletic motor skills 24–33; motor memory 24–9; and practice 30; and precision 30–3; and specialization 30

attention: definition of 63–4; distraction and 63–4; neuroanatomy of attentional networks 64–6; spatial 64; vigilance 66–9
attentional deficit-hyperactivity disorder (ADHD) 50
attentional networks, neuroanatomy of 64–6
axon 1

Baclofen 67
balance 74–6; flocculonodular lobe and 9–10; vestibular system and 74
balance sports 11
Bálint, R. 6, 20, 72
ball sports 11
basal ganglia 44
behaviors, goal-oriented 44, 51
benzodiazepines 67
Bohlhalter, S. 23
boxers, and TBI 85
bradykinesia 45
brain 40; damage 41; left hemisphere of 17–18; major lobes of 3; prefrontal cortex 8; programming movements and 40; and sports 11–12
brain anatomy 1–12; brief overview 1–3; cerebellum 9–11; motor and premotor cortex 7–8; prefrontal cortex 8; sensory cortex and sensory association areas 4–7; subcortical areas 8–9
Broca, Paul 16, 56
Bucy, Paul 31

cardiac muscles 38
catechol-O-methyltransferase (COMT) 80
cell body 1
cerebellum 9–11
cerebral herniation 82
cerebral peduncle 32

Chambers, R. A. 49
chronic traumatic encephalopathy 85–6
combat sports 11
computer tomography (CT) 82
concussion, defined 81
contralateral hemispheric specialization 61
Coombes, S. A. 68
Corbetta, M. 64, 69
Corkin, Suzanne 25
corpus callosum 8, 16, 57
corticospinal tract 13
Cotzias, G. 29

Darwin, Charles 62
defective brain development 41
defective response inhibition 83
deft movements 31
dementia, and exercise 87–8
dendrites 1
Denny-Brown, D. 49
depression 79–80; exercise and 79–80; medications 79; psychotherapy in 79; signs and symptoms 79; sports and 79–80
developmental dyslexia 42
directional akinesia 42
dopamine 28–9
dorsal "where" visual networks 6, 20
Douglas, Mike 28
Dubois, B. 52
dysarthria 85
dysmetria 11

electroencephalogram (EEG) 67
emotions 77–80, 84–5; anger 77–9; depression 79–80; and TBI 84–5
endorphins 80
endurance 37
epidural hemorrhage 82
episodic memory 82, 87
Erickson, K. I. 87
executive functions 83–4
exercise: and age 87–8; and dementia 87–8

First World War 43
Fisher, C. Miller 50
"flawed perseveration" 51
flocculonodular lobe 10
Fogassi, L. 32
force 37
frontal lobe: and action sequencing 33; injuries 43

functional magnetic resonance imaging (fMRI) 23
fusiform gyrus 34

Gage, Phineas 43
gamma-aminobutyric acid (GABA) 67
Geneva Convention 39
Geschwind, Norman 16–17, 31, 57, 59
global akinesia 42
goal-oriented behaviors 44, 51
Gonzalez-Rothi, Leslie 26
grasp reflex 50
"gray matter" 8

handedness (hand-arm preference) 53–62; contralateral hemispheric specialization and 61; genetics and 62; left-handedness 53–5, 58, 62
Hanna-Pladdy 31
Harlow, John 43
Hebb, Donald 27
Heilman, K. M. 31
hemispatial akinesia 42
Hillis, A. 65
hippocampus 24
Hitler, Adolf 47
Hogan, Ben 28
"how" programs 13–23
hypobulia 43
hypokinesia 45–6; bradykinesia and 45; defined 45

ideational apraxia 37
ideomotor apraxia 15–18, 58; and loss of movement representations 26
impaired gesture-pantomime discrimination and recognition 35–6
impersistence 83; see also motor (action) impersistence
inferior parietal lobe 19–20
internal medicine 40

Kaplan, Edith 57
Kertesz, Andrew 51
King, A. P. 27
Kissinger, Henry 27–8
Klein, D. 82
Kleist, Karl 43
Kümmel, J. 76
Kuypers, H. G. J. M. 32

language laterality-handedness hypothesis 57
Lawrence, D. G. 32

left hemisphere of brain 17–18; injury and motor acquisition task 25; and right handed people 25
Librium 67
Lichtheim, L. 36
Liepmann, Hugo 15, 16, 17–18, 31, 58, 59
limb akinesia 42
limbic system 8
"lower motor neurons" 32
Luria, A. R. 33, 51–2

magnetic resonance imaging (MRI) 82
Mantle, Mickey 54
Mark, Vernon 78
Marler, Peter 27
Martland, Harrison 85
McKee, A. C. 85–6
medial superior temporal (MST) 72–3
"melokinetic" (limb-kinetic) apraxia 33
memory 82–3; contusion and 82; episodic 82, 87; head injury and 82; loss of 82
Meyrignac, C. 33
middle temporal (MT) 72–3
Milner, Brenda 82
Milner, Peter 46
Mishkin, Mort 5, 20, 36, 71
modality-specific sensory association cortices 44
monoamine oxidase inhibitors 79
mood 77–80, 84–5; *see also* emotions
morphine 80
motor cortex 7–8, 13, 14, 60; and muscles 38
motor (action) impersistence 50–1; *see also* impersistence
motor learning, and age 27
motor memory, and athletic motor skills 24–9
motor perseveration: action-intention 51–2; flawed perseveration 51; recurrent perseveration 51; terminal 51
"movement formulas" 17
Murray, Eddie 54
muscles: cardiac 38; and motor cortex 38; skeletal 38; smooth 38
myelin 46

National Football League 85
Nauta, Wally 44
neuroanatomy of attentional networks 64–6
neurological impairments 41

neurology 40
neurons 1–3
neurotransmitters 2–3
New England Journal of Medicine 29
Nicklaus, Jack 28
norepinephrine 79–80
norepinephrine uptake inhibitors 79
nucleus accumbens 46, 47

occipital cortex 19–20
Ochipa, C. 26
Oldenziel, K. 30
Olds, James 46
optic ataxia 6, 20
optic chiasm 70

Palmer, Arnold 28
pantomime agnosia 36
parietal lobe: inferior 19–20; superior 20
Parkinson's disease 9, 42, 44–5, 47; and aging 29; and motor skills 28
pedunculopontine nucleus 76
perception: endorphins and 80; visional 70–3
persistence: action-intention 40; motor 40, 50–1; triage of patients and 40
phrenology 56
planning, and action-intention 45
Poling, A. 54
polymodal cortex 6, 20; *see also* inferior parietal lobe
positron emission tomography (PET) imaging 22
practice, and athletic motor skills 30
praxis system 13–23; and sports 13–23
precision, and athletic motor skills 30–3
prefrontal cortex 8, 64
premotor cortex 7–8; areas of 21
primary auditory cortex 4, 5
primary sensory cortex 4
primary visual cortex 4–5
prosopagnosia 34
"Punch Drunk" 85

"recurrent perseveration" 51
reduction laziness 43
reinforcement: random 49; reward 47–9
response inhibition, and action-intention 49–50
reticular activating system 81

reward: action-intention 46–9; network 46–7; reinforcement 47–9
Ritalin 67
Rose, Pete 54
rotary pursuit 25
Rothi, L. J. G. 36

Sandson, J. 51
Schwartz, R. L. 36
Scoville, W. B. 82
sensory association areas 4–7
sensory cortex 4–7
supplementary motor area (SMA) 7, 21
serotonin reuptake inhibitors 79
Shulman, G. L. 64, 69
skeletal muscles 38
Skinner, B. F. 49
SMA 33
smooth muscles 38
Soma (muscle relaxant) 67
somatosensory cortex 4–5
spatial attention 64
specialization, and athletic motor skills 30
speed sports 11
sports: accuracy 11; and action recognition 34–7; balance 11; ball 11; and brain 12; combat 11; depression and 79–80; and praxis system 13–23; speed 11; strength 11, 37–8; and TBI 84–5; *see also specific sports*
Sports Illustrated 85
Star Excursion Balance Test (SEBT) 75–6
strength: and endurance 37; and force 37
strength sports 11, 37–8
subcortical areas 8–9
substantia nigra 46, 48
superior longitudinal fasciculus 21–2
superior parietal lobe 20; *see also* parietal lobe
supramarginal gyrus 34, 59
synapse 1–2

Tebow, Tim 42
"terminal motor perseveration" 51
thalamus 9
Thorpe, William 27
"time space form picture of the movement" 17
traumatic brain injury (TBI) 81–6; and emotions 84–5; emotions and mood 84–5; executive functions 83–4; memory 82–3; and mood 84–5; and sports 84–5
traumatic injury 81–2
triage of patients 39–40; action inhibition 40; action initiation 40; action termination 41; persistence 40
tricyclic antidepressants 79

Ungerleider, Leslie 5, 19, 36, 71
University of Florida 42

Valium 67
ventral tegmental area 47, 48
ventral "what" visual networks 6
vermis 9
Verstichel, P 33
Vietnam War 39
vigilance, in attention 66–9
visional perception 70–3
visual agnosia 5

Watson, Bob 16
Wernicke, Karl 41, 56
Wernicke's area 41
West, M. J. 27
"What" stream 5
"Where" stream 6
"white matter" 8
Woods, Earl 28
Woods, Tiger 28
Wyke, Maria 25

Xanax 67

Yamashita, H. 25
Yerkes-Dobson curve 68

For Product Safety Concerns and Information please contact our EU
representative GPSR@taylorandfrancis.com
Taylor & Francis Verlag GmbH, Kaufingerstraße 24, 80331 München, Germany

www.ingramcontent.com/pod-product-compliance
Lightning Source LLC
Chambersburg PA
CBHW071823230426
43670CB00013B/2548